T0330991

2012
RECENT ADVANCES IN FINANCIAL ENGINEERING

Proceedings of the
International Workshop
on Finance 2012

2012
RECENT ADVANCES IN FINANCIAL ENGINEERING

Proceedings of the
International Workshop
on Finance 2012

The University of Tokyo, Japan
30–31 October 2012

editors

Akihiko Takahashi
The University of Tokyo, Japan

Yukio Muromachi
Tokyo Metropolitan University, Japan

Takashi Shibata
Tokyo Metropolitan University, Japan

World Scientific

NEW JERSEY · LONDON · SINGAPORE · BEIJING · SHANGHAI · HONG KONG · TAIPEI · CHENNAI

Published by

World Scientific Publishing Co. Pte. Ltd.

5 Toh Tuck Link, Singapore 596224

USA office: 27 Warren Street, Suite 401-402, Hackensack, NJ 07601

UK office: 57 Shelton Street, Covent Garden, London WC2H 9HE

Library of Congress Cataloging-in-Publication Data
International Workshop on Finance (2012 : Tokyo, Japan)
    Recent advances in financial engineering 2012 : proceedings of the International Workshop on
Finance 2012, the University of Tokyo, Japan, 30-31 October 2012 / editors Akihiko Takahashi,
The University of Tokyo, Japan, Yukio Muromachi, Tokyo Metropolitan University, Japan, Takashi
Shibata, Tokyo Metropolitan University, Japan.
        pages cm
    ISBN 978-9814571630 (hardcover : alk. paper) -- ISBN 9814571636 (hardcover : alk. paper)
    1. Financial engineering--Congresses. I. Takahashi, Akihiko. II. Muromachi, Yukio. III. Shibata,
Takashi (Associate professor) IV. Title.
    HG176.7.I575 2012
    332.64'5--dc23

    2013048375

British Library Cataloguing-in-Publication Data
A catalogue record for this book is available from the British Library.

In-house Editor: Philly Lim

Printed in Singapore

Preface

This book contains the Proceedings of the *International Workshop on Finance 2012*, which was held at the University of Tokyo on October 30 and 31, 2012. This workshop was organized by the Center for Advanced Research in Finance (CARF), Graduate School of Economics, the University of Tokyo, and Graduate School of Social Sciences, Tokyo Metropolitan University (TMU).

This annual workshop, which was first held in 2011, is a successor to the Daiwa International Workshop (2004 to 2008) and the KIER-TMU International Workshop (2009 to 2010). The workshop was designed for the exchange of new ideas in financial engineering and to serves as a bridge between academic researchers and practitioners. To these ends, the speakers shared various interesting ideas, information on new methods, and their up-to-date research results. In the 2012 workshop, we invited nine leading scholars, including three keynote speakers, from various countries, and the two-day workshop resulted in many fruitful discussions.

This book consists of eight papers, all refereed, that were related to the presentations at the International Workshop on Finance 2012. In these papers, the latest concepts, methods, and techniques related to current topics in financial engineering are proposed and reviewed. The editors hope that readers of this book will learn a great deal about cutting-edge research in financial engineering.

We would like to express our utmost gratitude to those who contributed their papers. Many thanks to all the anonymous reviewers who evaluated the papers submitted. Financial support from the Norinchukin Bank, the Global COE Program "The Research and Training Center for New Development in Mathematics," the Japan Society for Promotion of Science (JSPS) KAKENHI Grant Number 21241040, the TMU Program for Enhancing the Quality of University Education, and the Credit Pricing Corporation is greatly appreciated.

We are grateful to Mr. Satoshi Kanai for editing the manuscripts and to Ms. Alisha Nguyen, Ms. Agnes Ng, and the editorial committee members of World Scientific Publishing Co. for their kind assistance and support in publishing this book.

November 2013

Akihiko Takahashi, The University of Tokyo
Yukio Muromachi, Tokyo Metropolitan University
Takashi Shibata, Tokyo Metropolitan University

International Workshop on Finance 2012

Date

October 30 and 31, 2012

Place

The University of Tokyo, Tokyo, Japan

Organizer

Center for Advanced Research in Finance (CARF), Graduate School of Economics, The University of Tokyo

Graduate School of Social Sciences, Tokyo Metropolitan University

Supported by

The Norinchukin Bank

Global COE Program "The Research and Training Center for New Development in Mathematics"

Japan Society for Promotion of Science (JSPS) KAKENHI Grant Number 21241040

TMU Program for Enhancing the Quality of University Education

Credit Pricing Corporation

Program Committee

Tomio Arai, The University of Tokyo, Chair

Akihiko Takahashi, The University of Tokyo

Masaaki Fujii, The University of Tokyo

Yukio Muromachi, Tokyo Metropolitan University

Takashi Shibata, Tokyo Metropolitan University

Kensuke Ishitani, Tokyo Metropolitan University

Program

October, 30 (Tuesday)

Opening Address

    Tomio Arai (The University of Tokyo)

  Chair: Masaaki Fujii

10:00–11:00 Stéphane Crépey (Evry University)
    Plenary talk: Bilateral Counterparty Risk under Funding Constraints

11:00–11:45 Areski Cousin (University of Lyon)
    Dynamic Hedging of Portfolio Credit Risk in a Markov Copula Model

11:45–13:30, Lunch

  Chair: Kazutoshi Yamazaki

13:30–14:30 Alexander Lipton (Imperial College London)
    Plenary talk: Asymptotics for Exponential Levy Processes and Their Volatility Smile

14:30–15:15 Julien Grepat (University of Franche-Comte)
    On the Limit Behavior of Option Hedging Sets under Transaction Costs

15:15–15:45, Afternoon Coffee

  Chair: Lei Shi

15:45–16:30 Chun Ming Tam (Tokyo Metropolitan University)
    Fractional Brownian Motions in Financial Models and their Monte Carlo Simulation

October, 31 (Wednesday)

Chair: Sheung Chi Phillip Yam

10:00–11:00 Stéphane Crépey (Evry University)
Plenary talk: Dynamized Copulas and Applications to Counterparty Risk on Credit Derivatives

11:00–11:45 Elisabeth Van Laere (National University of Singapore)
Why Banks Disappear: A Forward Intensity Model for Default and Distressed Exits

11:45–13:30, Lunch

Chair: Pawel Zaczkowski

13:30–14:30 Fred Espen Benth (Oslo University)
Plenary talk: A General Approach to Pricing in Energy and Weather Markets

14:30–15:15 Sheung Chi Phillip Yam (The Chinese University of Hong Kong)
Linear Quadratic Mean Field Games

15:15–15:45, Afternoon Coffee

Chair: Kensuke Ishitani

15:45–16:30 Pawel Zaczkowski (The University of Cambridge)
Firms, Banks and Households

16:30–17:15 Lei Shi (University of Technology, Sydney)
Are You Smart Enough to Beat the Market?

Closing Address
Yukio Muromachi (Tokyo Metropolitan University)

Contents

Forward Prices in Markets Driven by Continuous-time Autoregressive Processes[*]

Fred Espen Benth and Sara Ana Solanilla Blanco

Centre of Mathematics for Applications,
Department of Mathematics, University of Oslo,
P.O. Box 1053, Blindern N-0316 Oslo, Norway
Email: fredb@math.uio.no   s.a.s.blanco@cma.uio.no

We analyse the forward price dynamics for contracts written on a spot following a continuous-time autoregressive dynamics. Prime examples of such spots could be power or freight rates, or weather variables like temperature and wind speed. It is shown that the forward price evolves according to template term structure functions, which are scaled by the deseasonalized spot and its derivatives. These template term structure functions can be expressed as a series of exponentially decaying functions with rates given by the (real parts) of the eigenvalues of the autoregressive dynamics. Moreover, the continuous-time autoregressive spot dynamics is differentiable up to an order less than the autoregressive order, and this is precisely the derivatives needed in the representation. The template term structures may produce humps in the forward curve. We consider several empirical examples for illustration based on a model relevant for the temperature market. A particular result of our analysis is that the paths of the forward price are non-differentiable, although the underlying spot is smooth. Our results offer insight into the dynamics of forward and futures prices for contracts in the markets for weather, shipping and power.

Key words: continuous AR processes, forward price, spot-forward relationship, weather markets, energy markets, interest rate theory, Lévy processes

---

[*]Send all correspondence to Fred Espen Benth, Centre of Mathematics for Applications, University of Oslo, P.O. Box 1053, Blindern N-0316 Oslo, Norway. Email: fredb@math.uio.no

1. Introduction

Continuous-time autoregressive, and more generally, continuous-time autoregressive moving average processes have in recent years become popular in various financial applications. These non-Markovian processes allow for memory effects in the dynamics, and have been succesfully used in the modelling of different weather variables (see e.g. Benth and Šaltytė Benth [8]), interest rates (see Zakamouline et al. [22]) and commodity prices like power, freight rates or even oil (see e.g., Garcia et al. [15], Benth et al. [6] and Paschke and Prokopczuk [19]). The purpose of this paper is to investigate the forward price dynamics for contracts written on underlying spots following a continuous-time autoregressive dynamics. Such spots may be classical assets like a commodity, but also include temperature and wind, and even power or freight rates. Typically, all these examples are spots which cannot be traded in a portfolio sense.

If an underlying spot of a forward contract cannot be traded liquidly, like for example power, the classical spot-forward relationship based on the buy-and-hold hedging strategy breaks down. We resort to the no-arbitrage theory, and demand that the forward price is defined as the conditional expected spot price at delivery of the contracts, with the expectation taken under some pricing measure. This ensures that the forward price dynamics is a martingale under the pricing measure. Using the Esscher transform to construct a class of pricing measures (being simply a parametric class of equivalent probabilities), we can compute analytically the forward price dynamics. We show that this price dynamics is explicitly represented in terms of the spot and its derivatives up to order $p - 1$, with p being the autoregressive order.

We consider general Lévy-driven continuous-time autoregressive processes of order $p \geq 1$ as defined by Brockwell [11]. The continuous-time autoregressive processes are semimartingales. In fact, they have differentiable paths of order up to $p - 1$, where the dynamics of the derivatives can be explicitly stated. Of major importance in our analysis is the representation of the continuous-time autoregressive process as a p-dimensional Ornstein-Uhlenbeck process driven by the same (one-dimensional) Lévy process. The stationarity of the continuous-time autoregessive process is ensured by the drift matrix in this Ornstein-Uhlenbeck process having eigenvalues with negative real part. This matrix and its eigenvalues are crucial in understanding the forward price dynamics.

We compute the forward price dynamics, and show that it can be decomposed into a deterministic and stochastic part. The deterministic part includes possible seasonality structures and the market price of risk, where the latter comes from the parametric choice of pricing measures. In this paper, our concern is on the stochastic part of the forward dynamics, which is expressable as a sum of p functions in *time to maturity* x, denoted $f_i(x)$, $i = 1, \ldots, p$, where $f_i(x)$ is scaled by the $i - 1$th derivative of this continuous-time autoregressive process. These derivatives are directly linked to derivatives of (some function) of the underlying spot dynamics. We call $f_i(x)$ template forward term structure functions, as they are the basic building blocks for the forward curve. These templates can make up forward curves in contango or backwardation, including humps of different sizes. The humps will occur when the slope (first derivative) and/or curvature (second derviative) of the spot is particularly big. We also include an analysis of forwards written on the average of the underlying spot, being relevant in weather and power markets as their contracts have a delivery *period* rather than a delivery *time*.

Due to stationarity of the spot model, forward prices far from maturity of the contract will be essentially non-stochastic (constant). However, close to maturity they will start to vary stochastically according to the size of the spot and its derivatives. Noteworthy is that the forward dynamics is much more erratic than the underlying spot, due to the fact that it depends on higher-order derivatives of the spot, which eventually have paths similar to the driving Lévy process. The spot, on the other hand, is smooth in the sense of being differentiable up to some order.

We illustrate our results by numerical examples. These examples are based on the empirical estimation of a continuous-time autoregressive model of order 3 on daily temperature data observed over more than 40 years in Stockholm, Sweden. Apart from presenting different shapes of the forward curve and its dynamics, we also present how to apply our results to recover the derivatives from an observed path of spot prices (or, temperatures). Natural in our context is to resort to finite differencing of the continuous-time autoregressive process observed discretely, which turns out to give a reasonable approximation.

Our results are presented as follows: in the next Section we define continuous-time autoregressive processes and present some applications of these in different financial contexts. Sections 3 and 4 contain our main results, with the derivation of the forward prices and analysis of the forward curve as a function of time to delivery. Finally, we conclude and make an outlook.

2. Continuous-time Autoregressive Processes

In this Section we introduce the class of continuous-time autoregressive processes proposed by Doob [13] and later intensively studied by Brockwell [11].

Let $(\Omega, \mathcal{F}, P)$ be a probability space equipped with a filtration $\{\mathcal{F}_t\}_{t \in \mathbb{R}}$ satisfying the usual conditions (see e.g. Karatzas and Shreve [18]). We assume L is a real-valued two-sided square-integrable Lévy process, and choose to work with the RCLL version (right-continuous, with left-limits). Denote by $\mathbf{e}_k$ the kth canonical basis vector in $\mathbb{R}^p$ for $p \in \mathbb{N}$ and $k = 1, \ldots, p$. We define the *continuous–time autoregressive process* of order p (from now on, a CAR(p)-process) to be

$$(1) \qquad Y(t) = \int_{-\infty}^{t} \mathbf{e}_1^{\mathrm{T}} e^{A(t-s)} \mathbf{e}_p \, dL(s),$$

for $t \geq 0$, whenever the stochastic integral makes sense. Here, $\mathbf{x}^{\mathrm{T}}$ denotes the transpose of a vector or matrix $\mathbf{x}$, and A is the $p \times p$ matrix

$$(2) \qquad A = \begin{pmatrix} \mathbf{0}_{p-1 \times 1} & I_{p-1} \\ -\alpha_{p\cdots} & \cdots -\alpha_1 \end{pmatrix}$$

for positive constants α_i, $i = 1, \ldots, p$. The expression $\exp(At)$ is interpreted as the matrix exponential for any time $t \geq 0$. In the next lemma we derive the characteristic function of the CAR(p)-process $Y(t)$ and show that it is stationary:

Lemma 2.1. *Let A have eigenvalues with negative real part. For any $x \in \mathbb{R}$ and $t \geq 0$ it holds that*

$$\ln \mathbb{E}\left[e^{ixY(t)} \right] = \int_0^\infty \psi_L\left(x\mathbf{e}_1^{\mathrm{T}} e^{As} \mathbf{e}_p \right) ds \,,$$

where $\psi_L(x) = \ln \mathbb{E}[\exp(ixL(1))]$ is the characteristic exponent of L.

Proof. In the proof, we consider the process

$$\widetilde{Y}(t) = \int_0^t \mathbf{e}_1^{\mathrm{T}} e^{A(t-s)} \mathbf{e}_p \, dL(s) \,,$$

and we show that this has a stationary limit as time t tends to infinity. Let $\{t_j\}_{j=1}^n$ be a sequence of nested partitions of the interval $[0, t]$. From the independent increment property of the Lévy process we find for $x \in \mathbb{R}$ and the definition of its characteristic function,

$$(3) \qquad \mathbb{E}\left[e^{ix\widetilde{Y}(t)} \right] = \lim_{n \to \infty} \mathbb{E}\left[\prod_{j=1}^n e^{i[xg(t-t_j)]\Delta L(t_j)} \right]$$

$$= \lim_{n \to \infty} \sum_{j=1}^n \psi_L(xg(t - t_j))\Delta t_j$$

$$= \exp\left(\int_0^t \psi_L(xg(t - s)) \, ds \right)$$

$$= \exp\left(\int_0^t \psi_L(xg(s)) \, ds \right).$$

where $\Delta L(t_j) = L(t_{j+1}) - L(t_j)$, $\Delta t_j = t_{j+1} - t_j$, and where we have used the shorthand notation $g(u) = \mathbf{e}_1^{\mathrm{T}} e^{Au} \mathbf{e}_p$.

As the eigenvalues of A have negative real part, we can express $g(s) = \mathbf{e}_1^{\mathrm{T}} e^{As} \mathbf{e}_p$ as a sum of exponentials scaled by trigonometric functions. The exponentials decay at rates given by the negative real part of the eigenvalues. Hence, we can majorize $g(s)$ as

$$|g(s)| \leq C e^{\mathrm{Re}(\lambda_1)s} \,,$$

where λ_1 is the eigenvalue with smallest absolute value of the real part. Finally we take logarithm in (3) and apply Theorem 17.5 of Sato [21] to conclude that

$$\lim_{t \to \infty} \int_0^t \psi_L(xg(s)) \, ds = \int_0^\infty \psi_L(xg(s)) \, ds.$$

This concludes the proof. □

From now on we assume that A has eigenvalues with negative real part, which by the above lemma makes $Y(t)$ in (1) well-defined and stationary.

A particular case of interest is when $L(t) = \sigma B(t)$, for $\sigma > 0$ a constant and B a Brownian motion. Then the characteristic exponent of L is $\psi_L(x) = -\sigma^2 x^2 / 2$, and the distribution of Y has characteristic exponent

$$\int_0^\infty \psi_L \left(x \mathbf{e}_1^T e^{As} \mathbf{e}_p \right) ds = -\frac{1}{2} \sigma^2 x^2 \int_0^\infty \mathbf{e}_1^T e^{As} \mathbf{e}_p \mathbf{e}_p^T e^{A^T s} \mathbf{e}_1 \, ds$$

Hence, it becomes normally distributed, with zero mean and

$$\sigma^2 \int_0^\infty \mathbf{e}_1^T e^{As} \mathbf{e}_p \mathbf{e}_p^T e^{A^T s} \mathbf{e}_1 \, ds$$

being the variance.

CAR(p)-processes can be thought of as a subclass of the so-called *continuous-time autoregressive moving average* processes, denoted CARMA(p, q) with $p, q \in \mathbb{N}$ and $0 \le q < p$. The order of the autoregressive part is given by p and the order of the moving average part is given by q. CARMA(p, q)-processes are defined as,

$$Y(t) = \int_{-\infty}^t \mathbf{b}^T e^{A(t-s)} \mathbf{e}_p \, dL(s),$$

for a vector $\mathbf{b} \in \mathbb{R}^p$ with the property

$$\mathbf{b}^T = \begin{pmatrix} b_0 & b_1 & \cdots & b_{q-1} & 1 & 0 & \cdots & 0 \end{pmatrix}.$$

We obviously recover the CAR(p)-process by letting $\mathbf{b} = \mathbf{e}_1$. CAR($p$), or CARMA($p, q$), processes are members of the much more general class of Lévy semistationary (LSS) processes

(4) $$X(t) = \int_{-\infty}^t g(t - s)\sigma(s) \, dL(s),$$

where $g : \mathbb{R}_+ \mapsto \mathbb{R}$ is a deterministic function and $\sigma(t)$ a predictable stochastic process such that

(5) $$\mathbb{E}\left[\int_{-\infty}^t g^2(t - s)\sigma^2(s) \, ds \right] < \infty,$$

for all $t \geq 0$. This makes the stochastic integral in the definition of the LSS-process well-defined in the sense of stochastic integration with respect to semimartingales, as defined in Protter [20]. The process $\sigma(t)$ is usually interpreted as a stochastic volatility or intermittency, see [1]. In the situation of a CAR(p)-process Y, we see that

$$g(t) = \mathbf{e}_1^T e^{At} \mathbf{e}_p,$$

for $t \geq 0$ and $\sigma(t) = 1$. It is of course not a problem to include a stochastic volatility in the definition (1) of the CAR(p)-process, however, we shall not do so here for the sake of simplicity. It is simple to see that for this particular g, the integrability condition (5) becomes

$$\int_0^\infty g^2(s)\,ds < \infty$$

which holds by estimating g using matrix norms.

We note that for $p = 1$, the matrix A collapses into $A = -\alpha_1$, and we find $g(t) = \exp(-\alpha_1 t)$. For this particular case we recognize $Y(t)$ as an Ornstein-Uhlenbeck process. In fact, general CAR(p)-processes can be defined via multivariate Ornstein-Uhlenbeck processes, as we show now: Introduce the $\mathbb{R}^p$-valued stochastic process $\mathbf{X}$ as the solution of the linear stochastic differential equation

$$(6) \qquad\qquad d\mathbf{X}(t) = A\mathbf{X}\,dt + \mathbf{e}_p\,dL(t).$$

The stationary solution of this multivariate Ornstein-Uhlenbeck process is (using Itô's Formula for jump processes, see Ikeda and Watanabe [17]),

$$(7) \qquad\qquad \mathbf{X}(t) = \int_{-\infty}^t e^{A(t-s)} \mathbf{e}_p\,dL(s).$$

We see that

$$(8) \qquad\qquad Y(t) = \mathbf{e}_1^T \mathbf{X}(t).$$

Hence, the CAR(p)-process is the first coordinate process of the multivariate Ornstein-Uhlenbeck process $\mathbf{X}(t)$, which is a real-valued Ornstein-Uhlenbeck process for $p = 1$ as already observed.

The paths $t \mapsto Y(t)$ of the CAR(p)-process will be smooth. This is a consequence of the next lemma:

Lemma 2.2. *Let $p > 1$. It holds that*

$$Y(t) = \int_0^t \int_{-\infty}^s \mathbf{e}_2^T e^{A(s-u)} \mathbf{e}_p\,dL(u)\,ds,$$

for $t \geq 0$.

Proof. From Basse and Pedersen [2] (see also Proposition 3.2 of Benth and Eyjolfsson [4] for the similar statement) we find that an LSS process $X(t)$ as in (4) with absolutely continuous kernel function g with a derivative satisfying $\int_0^\infty g'(s)^2\,ds < \infty$ and $|g(0)| < \infty$ has the representation

$$dX(t) = \int_{-\infty}^t g'(t-s)\sigma(s)\,dL(s)\,dt + g(0)\sigma(t)\,dL(t)\,,$$

for $t \geq 0$. For the case of a CAR(p)-process, we observe that $g(t) = \mathbf{e}_1^{\mathsf{T}}\exp(At)\mathbf{e}_p$ is continuously differentiable with $g(0) = \mathbf{e}_1^{\mathsf{T}}\mathbf{e}_p = 0$ and

$$g'(t) = \mathbf{e}_1^{\mathsf{T}}Ae^{At}\mathbf{e}_p\,.$$

By the definition of A in (2), we see that $\mathbf{e}_1^{\mathsf{T}}A = \mathbf{e}_2^{\mathsf{T}}$. But as the eigenvalues of the matrix A have negative real part, we find that $\int_0^\infty g'(s)^2\,ds < \infty$ as $|g'(s)|$ will be bounded by an exponentially decaying function. The result follows. $\qquad\square$

This shows in particular that the process $Y(t)$ is of finite variation. Moreover, we see that the derivative of $Y(t)$ exists, and is equal to

$$Y'(t) = \int_{-\infty}^t \mathbf{e}_2^{\mathsf{T}}e^{A(t-s)}\mathbf{e}_p\,dL(s)\,.$$

In fact, we can iterate the proof of Lemma 2.2 using the definition of A in (2) to show that the following smoothness result holds for $Y(t)$.

Proposition 2.1. *Let Y be a CAR(p)-process for $p > 1$. Then the paths $t \mapsto Y(t)$ for $t \geq 0$ are $p-1$ times differentiable, with ith derivative, $Y^{(i)}(t)$, given by*

$$Y^{(i)}(t) = \int_{-\infty}^t \mathbf{e}_{i+1}^{\mathsf{T}}e^{A(t-s)}\mathbf{e}_p\,dL(s)\,,$$

for $i = 1,\ldots,p-1$.

Note that $Y(t)$ is not a Markovian process. However, due to the representation via $\mathbf{X}$ above, it can be viewed as a p-dimensional Markovian process. We remark that the smoothness property and the representation of the derivatives of $Y(t)$ could also be proven by resorting to the stochastic differential equation for $\mathbf{X}(t)$.

We next turn our attention to some of the applications of CAR(p)-processes. First, let us consider a model for the time dynamics of temperature in a specific location. In Figure 1 we see the daily average temperatures in Vilnius, Lithuania, a city located in north-east Europe. The figure shows the average value of the maximum and minimum temperature recorded on each day, ranging over a five year period. The temperatures are measured in degrees Celsius (°C).

8

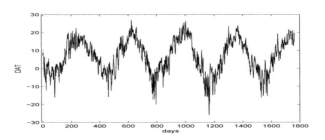

Figure 1. Five years of daily average temperatures measured in Vilnius.

In Benth and Šaltytė Benth [8] it is shown that an appropriate model for the time dynamics of these temperatures is given by

$$(9) \qquad\qquad T(t) = \Lambda(t) + Y(t),$$

where $T(t)$ is the temperature at time $t \geq 0$, $\Lambda(t)$ is some deterministic function modelling the seasonally varying mean level, and $Y(t)$ is a CAR(p)-process as above. For Vilnius, a seasonal mean function can be chosen as

$$\Lambda(t) = a_0 + a_1 t + a_2 \sin(2\pi(t - a_3)/365),$$

i.e., a yearly cycle of amplitude a_2, shifted by a_3, and a linear trend $a_0 + a_1 t$. The level is a_0, and a_1 indicates an increasing average temperature over time that may be attributed to urbanization and climate change. Furthermore, a statistical analysis of the deseasonalized temperatures $T(t) - \Lambda(t)$ reveals an autoregressive structure of order $p = 3$, being stationary. Hence, a CAR(3)-process $Y(t)$ is appropriate. Moreover, the analysis in Benth and Šaltytė Benth [8] reveals that $L(t) = \sigma B(t)$, B being a Brownian motion. In fact, a time-dependent volatility $\sigma(t)$ is also proposed to capture the seasonal variance observed in the data.

Another weather variable that can be conveniently modelled by CAR(p)-processes is wind speed. In Figure 2 we have plotted the wind speed in meters per second (m/s) measured in Vilnius at the same location as the temperature measurements discussed above.

Benth and Šaltytė Benth [8] propose an exponential model for the wind speed dynamics $W(t)$ given by

$$(10) \qquad\qquad W(t) = \exp(\Lambda(t) + Y(t)) .$$

Hence, the logarithm of the wind speed follows a seasonal mean function $\Lambda(t)$ and a CAR(p)-process. In the case of wind speeds, the seasonal mean function $\Lambda(t)$ is more complex than the simple sine-function with trend chosen for temperatures.

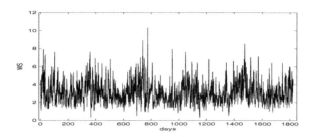

Figure 2. Five years of daily average wind speed measured in Vilnius.

However, it is also for wind speeds in Vilnius appropriate to choose Y to be a CAR(3)-process driven by a Brownian motion. Also in this case one observes a seasonally varying variance $\sigma(t)$.

It is to be noted that many studies have confirmed the CAR(3)-structure of temperature dynamics, see e.g. Härdle and Lopez-Cabrera [16] for analysis of German temperature data. Wind speeds at different locations vary between CAR(3) and CAR(4). For example, a study of wind speeds in New York by Benth and Šaltytė Benth [7] shows that CAR(4) is the best choice. In the empirical studies presented above, the stochastic part becomes stationary in the limit as the eigenvalues of the A matrix in the CAR(3)-process have negative real part. From a practical point of view, it is rather natural that wind and temperature are stationary phenomena around its mean level. These models for wind and temperature have been applied to weather derivatives pricing and hedging, in particular futures written on temperature and wind speed indexes, see Benth et al. [8] and Benth & S. Benth [9]. We will return to this, in the next Section.

Freight rates can be modelled using CAR(p)-processes. Benth et al. [6] perform an empirical study of the daily observed Baltic Capesize and Baltic Panamax Indexes, which are indexes created from assessments of 10, resp. 4, time charter rates of Capesize, resp. Panamax, vessels. Indeed, the dynamics of the logarithmic freight rates can in both cases be modelled by CAR(3)-processes, where the Lévy process is normal inverse Gaussian distributed. We have plotted the time evolution of the daily Baltic Capesize Index in Figure 3 over the period from March 1999 to November 2011.

We note that the Baltic Exchange in London, UK, organizes a trade in futures contracts written on the average of these indexes over given time periods.

Benth et al. [22] have shown that a Brownian-driven CARMA(2,1) model explains well the term structure of volatility for forward rates from UK treasury bonds. Another application area of CAR(p)-processes is commodity markets. Garcia et al. [15] demonstrated that deseasonalized electricity spot prices observed at the German power exchange EEX follow a CARMA(2,1) process driven by an

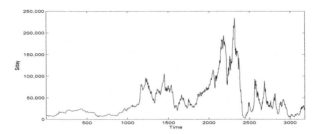

Figure 3. The Baltic Capesize Index over the period from March 1999 to November 2011.

α-stable Lévy process. A more recent study by Benth et al. [5] takes electricity futures prices into account as well, and extends the spot model into a two-factor CAR(1) and CARMA(2,1) dynamics. At the EEX, and other power exchanges, futures and forward contracts are traded settled on the average electricity spot price over a specific time period. In [19] Paschke and Prokopzcuk have proposed a Brownian-driven CARMA(2,1)-model for the time dynamics of crude oil, and estimated this to crude oil futures with the aim of studying various term structures. We will analyse futures pricing based on general CARMA-processes in another context, but mention these areas of applications as they are close to CAR-processes.

3. Forward Pricing

In this Section we consider the problem of deriving a forward price and how to represent this in terms of the underlying spot with a price dynamics given by $S(t)$. To this end, let $f(t, T)$ denote the forward price at time $t \geq 0$ of a contract delivering the underlying spot at time T. In classical financial theory, one resorts to the so-called buy-and-hold hedging strategy in the underlying asset to reach the spot-forward relationship

$$f(t, T) = S(t)e^{r(T-t)},$$

where $r > 0$ is the constant risk-free interest rate.

Considering forward contracts in markets for electricity, weather or freight, one cannot store the spot. There is no way one can buy and store temperature, nor wind, and by the very nature of power this is not storable either. The spot rates of freight are indexes, and hence cannot be used for hedging in a portfolio either, like the spot interest rate. Thus, in these markets we cannot resort to the hedging argument replicating a long forward position by holding a spot. This implies that the spot-forward relationship breaks down in the typical markets we have in mind.

From the arbitrage theory in mathematical finance (see e.g. Bingham and Kiesel [10]), we know that all tradeable assets in a market must have a martingale price dynamics under a risk neutral measure Q. In particular, forward prices must be martingales under some risk neutral probability Q. In a market where the spot cannot be traded (or stored), the probability Q does not need to be an equivalent martingale measure in the sense that the spot dynamics is a Q-martingale (after discounting). The only requirement is that Q is equivalent to P, and that we ensure the martingale property of the forward price. We refer to Q as a *pricing measure* rather than an equivalent martingale measure. As $f(T, T) = S(T)$, we get by such a martingale requirement that

$$(11) \qquad f(t, T) = \mathbb{E}_Q [S(T) | \mathcal{F}_t] .$$

To have $f(t, T)$ well-defined, we suppose that $S(t) \in L^1(Q)$ for all $t \in [0, T]$. The pricing measure Q plays here much of the same role as risk loading does in insurance, as we may view the forward as an insurance contract locking in the spot at delivery. We refer to Benth et al. [9] and Benth and Šaltytė Benth [8] for more on the relationship between spot and forwards in markets where the spot cannot be stored.

In the remaining part of this section we assume that the spot price dynamics is given by two possible models, arithmetic or geometric dynamics driven by a CAR(p)-process: In the arithmetic case, we assume

$$(12) \qquad S(t) = \Lambda(t) + Y(t) ,$$

with $\Lambda : \mathbb{R}_+ \mapsto \mathbb{R}$ being a deterministic measurable function assumed to be bounded on compacts. The process $Y(t)$ is the CAR(p) dynamics defined in (1). The geometric model is assumed to be

$$(13) \qquad S(t) = \exp(\Lambda(t) + Y(t)) .$$

We consider these two models in order to cover the models for wind, temperature and freight rates discussed above. We also have in mind applications to energy and commodity markets, as well as fixed-income theory, where also both arithmetic and geometric models are relevant. Note that we also include a deterministic function Λ to capture a possible seasonality. Although this will only play a technical role in the computations to follow, we include it for completeness.

It is worth mentioning that a forward contract on wind or temperature, or on power, delivers over a fixed period of time and not at a fixed point in time T. For example, a forward contract on power will typically deliver power continuously to the owner of the contract over an agreed period. On the EEX power exchange in Germany, such delivery periods can be specific weeks, months, quarters, and even years. Hence, buying a forward on power entitles you to the delivery of power over a period $[T_1, T_2]$, meaning that you receive

$$\int_{T_1}^{T_2} S(u) \, du ,$$

with S being the power spot price at time u. In the market, the delivery is settled financially, meaning that the owner receives the money-equivalent of the above. The forward price of power is denoted per MWh, so by definition it becomes

$$(14) \qquad F(t, T_1, T_2) = \mathbb{E}_Q \left[\frac{1}{T_2 - T_1} \int_{T_1}^{T_2} S(u) \, du \, | \, \mathcal{F}_t \right].$$

At the Chicago Mercantile Exchange (CME) in the US there is a market for forwards[1] written on temperature indexes measured in various cities world-wide. For example, forwards settled on the average temperature measured in Tokyo over given months can be traded. This will yield a temperature forward price as in (14), with $S(u)$ interpreted as the temperature in Tokyo at time u in the measurement period $[T_1, T_2]$. There are three other temperature indexes used for settlement of forwards in this weather market, called HDD, CDD and CAT. We refer to Benth and Šaltytė Benth [8] for a definition and analysis of these. Finally, in the freight market the forward is also settled on the average spot freight rates over a period in time. This yields then as well a forward price given by (11) using $S(u)$ as the spot freight rate.

Note that by the Fubini-Tonelli theorem (see e.g. Folland [14]), we find

$$(15) \qquad F(t, T_1, T_2) = \frac{1}{T_2 - T_1} \int_{T_1}^{T_2} \mathbb{E}_Q [S(u) | \mathcal{F}_t] \, du = \frac{1}{T_2 - T_1} \int_{T_1}^{T_2} f(t, u) \, du.$$

Therefore, to price forwards in markets for weather, power and freight, it is sufficient to analyse forward prices with fixed time of delivery $f(t, T)$, as the forwards settling over a time period is simply an average of these.

We must fix a pricing measure Q, and we do so by introducing a parametric family of probabilities Q_θ for $\theta \in \mathbb{R}$ given by the Esscher transform. Introduce the process Z

$$(16) \qquad Z(t) = \exp(\theta L(t) - \psi_L(-i\theta)t),$$

for $t \geq 0$. Observe that $\psi_L(-i\theta)$ is the logarithm of the moment generating function of L. To have this process Z well-defined, we assume that L has moments of exponential order, meaning that

$$\mathbb{E}(\exp(kL)) < +\infty, \quad k > 0.$$

We observe that Z is a martingale with $Z(0) = 1$. Define a probability measure such that the Radon-Nikodym derivative has density process $Z(t)$, that is,

$$(17) \qquad \frac{dQ_\theta}{dP} \bigg|_{\mathcal{F}_t} = Z(t).$$

---

[1] We will not distinguish between forwards and futures in this paper. The contracts on temperatures at CME are futures-style, whereas the power contracts mentioned earlier may be of forward and futures style.

Remark that we perform a measure change only for the part of the Lévy process living on the positive time $t \geq 0$. As L is two-sided, it is also defined for $t < 0$. However, this part is independent of L defined on $t \geq 0$. We do not make any change of measure for negative times. We can accomodate this by extending the definition of Z to be equal to one for all times $t < 0$.

From Benth and Šaltytė Benth [8], Proposition 8.3, one finds that the Lévy property of L is preserved under this change of measure, and the characteristic exponent of L with respect to Q_θ becomes

(18) $$\psi_{L,\theta}(x) = \psi_L(x - i\theta) - \psi_L(-i\theta).$$

It is easily seen that if the Lévy measure of L is $\ell(dz)$, then the Lévy measure under Q becomes $\exp(\theta z)\ell(dz)$, that is, the measure is exponentially tilted. One refers to the constant θ as the *market price of risk*. Remark that for the case $L = B$, we find that the Esscher transform coincides with the Girsanov transform. To see this, take into account that in this case $\psi_L(x) = -x^2/2$ and therefore

$$Z(t) = \exp\left(\theta B(t) - \frac{1}{2}\theta^2 t\right),$$

which we recall from Girsanov's Theorem (see e.g. Karatzas and Shreve [18]) to be the density of a measure Q_θ such that the process W_θ defined by

$$dW_\theta(t) = -\theta \, dt + dB(t),$$

is a Q_θ-Brownian motion. From (18) we see that the characteristic exponent of B with respect to Q_θ becomes $\theta x i - x^2/2$, in line with the above.

We find the forward price for the arithmetic and geometric spot price models: these results can be found in Benth et al. [9] and Benth and Šaltytė Benth [8], but we recall them here with a slightly modified proof for the convenience of the reader.

Proposition 3.1. *Let $S(t)$ have the dynamics given by (12). Then for a $\theta \in \mathbb{R}$ we have*

$$f(t, T) = \Lambda(T) + \mathbf{e}_1^{\mathsf{T}} e^{A(T-t)} \mathbf{X}(t) - i\psi_L'(-i\theta)\mathbf{e}_1^{\mathsf{T}} A^{-1} \left(e^{A(T-t)} - I\right) \mathbf{e}_p.$$

for $0 \leq t \leq T$ and with $\mathbf{X}(t)$ defined as in (7).

Proof. Fix $\theta \in \mathbb{R}$. Then, by definition of the forward price and the spot dynamics we find

$$f(t, T) = \mathbb{E}_{Q_\theta}[S(T) \,|\, \mathcal{F}_t] = \Lambda(T) + \mathbb{E}_{Q_\theta}[\int_{-\infty}^{T} \mathbf{e}_1^{\mathsf{T}} e^{A(T-s)} \mathbf{e}_p \, dL(s) \,|\, \mathcal{F}_t],$$

But from splitting the integral into two pieces, one ranging from $-\infty$ to t, and the other from t to T, we find using $\mathcal{F}_t$-adaptedness on the former and independence of increments of Lévy processes of the latter, that

(19) $$\mathbb{E}_{Q_\theta}\left[\int_{-\infty}^{T} \mathbf{e}_1^{\mathrm{T}} e^{A(T-s)} \mathbf{e}_p\, dL(s) \,|\, \mathcal{F}_t\right]$$

$$= \int_{-\infty}^{t} \mathbf{e}_1^{\mathrm{T}} e^{A(T-s)} \mathbf{e}_p\, dL(s) + \mathbb{E}_{Q_\theta}\left[\int_{t}^{T} \mathbf{e}_1^{\mathrm{T}} e^{A(T-s)} \mathbf{e}_p\, dL(s)\right].$$

For the expectation, we have that

$$\mathbb{E}_{Q_\theta}\left[\int_{t}^{T} \mathbf{e}_1^{\mathrm{T}} e^{A(T-s)} \mathbf{e}_p\, dL(s)\right]$$

$$= -i\frac{d}{dx}\mathbb{E}_{Q_\theta}\left[\exp\left(ix\int_{t}^{T} \mathbf{e}_1^{\mathrm{T}} e^{A(T-s)} \mathbf{e}_p\, dL(s)\right)\right]\Bigg|_{x=0}.$$

Adapting the first part of the proof of the Lemma 2.1 (working under Q_θ rather than the probability measure P), we find

$$\mathbb{E}_{Q_\theta}\left[\int_{t}^{T} \mathbf{e}_1^{\mathrm{T}} e^{A(T-s)} \mathbf{e}_p\, dL(s)\right] = -i\psi_{L,\theta}'(0)\int_{0}^{T-t} \mathbf{e}_1^{\mathrm{T}} e^{A(s)} \mathbf{e}_p\, ds.$$

since $\psi_{L,\theta}(0) = 0$. But since $\psi_{L,\theta}'(0) = \psi_L'(-i\theta)$, and integrating the matrix exponential, we get

$$\mathbb{E}_{Q_\theta}\left[\int_{t}^{T} \mathbf{e}_1^{\mathrm{T}} e^{A(T-s)} \mathbf{e}_p\, dL(s)\right] = -i\psi_L'(-i\theta)\mathbf{e}_1^{\mathrm{T}} A^{-1}\left(e^{A(T-t)} - I\right)\mathbf{e}_p.$$

This shows that last term of the forward price $f(t, T)$.

Let us now consider the first term on the right-hand side of (19). From the representation (8) we have that $Y(t) = \mathbf{e}_1^{\mathrm{T}} \mathbf{X}(t)$ with $\mathbf{X}(t)$ given in (7). From the stochastic differential equation (6) and Itô's Formula for jump processes (see Ikeda and Watanabe [17]), we find

$$\mathbf{X}(T) = e^{A(T-t)}\mathbf{X}(t) + \int_{t}^{T} e^{A(T-s)} \mathbf{e}_p\, dL(s).$$

But on the other hand we know that

$$\mathbf{X}(T) = \int_{-\infty}^{T} e^{A(T-s)} \mathbf{e}_p\, dL(s).$$

Hence, it follows that

$$\int_{-\infty}^{t} \mathbf{e}_1^{\mathrm{T}} e^{A(T-s)} \mathbf{e}_p\, dL(s) = \mathbf{e}_1^{\mathrm{T}} e^{A(T-t)}\mathbf{X}(t).$$

This proves the Proposition. □

We observe that the second term in the forward price dynamics is closely related to the spot price $S(t)$ at time t. The third term is appearing as a result of the introduction of a pricing measure. The case $\theta = 0$ would lead to a similar term with $\psi'_L(0)$ rather than $\psi'_L(-i\theta)$. Indeed, the θ measures the *risk premium* in the market since

$$f(t, T) - \mathbb{E}[S(T) \mid \mathcal{F}_t] = -i(\psi'_L(-i\theta) - \psi'_L(0))\mathbf{e}_1^{\mathsf{T}} A^{-1} \left(e^{A(T-t)} - I \right) \mathbf{e}_p.$$

By observing forward prices in the market in question, one can calibrate θ.

Let us state the forward price in case of a geometric spot price model:

Proposition 3.2. *Let $S(t)$ have the dynamics given by (13). Then for a $\theta \in \mathbb{R}$ we have*

$$\ln f(t, T) = \Lambda(T) + \mathbf{e}_1^{\mathsf{T}} e^{A(T-t)} \mathbf{X}(t) + \int_0^{T-t} \psi_{L,\theta}\left(-i\mathbf{e}_1^{\mathsf{T}} e^{As} \mathbf{e}_p\right) ds,$$

for $0 \le t \le T$ and with $\mathbf{X}(t)$ defined in (7).

Proof. Arguing as in the proof of Proposition 3.1 by independence and measurability of the Lévy process, we find

$$f(t, T) = \exp\left(\Lambda(T) + \int_{-\infty}^{t} \mathbf{e}_1^{\mathsf{T}} e^{A(T-s)} \mathbf{e}_p \, dL(s)\right)$$
$$\times \mathbb{E}_{Q_\theta}\left[\exp\left(\int_t^T \mathbf{e}_1^{\mathsf{T}} e^{A(T-s)} \mathbf{e}_p \, dL(s)\right)\right].$$

Adapting the proof of Lemma 2.1 we find

$$\mathbb{E}_{Q_\theta}\left[\exp\left(\int_t^T \mathbf{e}_1^{\mathsf{T}} e^{A(T-s)} \mathbf{e}_p \, dL(s)\right)\right] = \exp\left(\int_0^{T-t} \psi_{L,\theta}\left(-i\mathbf{e}_1^{\mathsf{T}} e^{A(s)} \mathbf{e}_p\right) ds\right).$$

The remaining part of the proof goes as for the proof of Proposition 3.1. □

If we are interested in the forward price for a contract which delivers the spot over a period, we recall the relationship (15) and see that for the arithmetic case one may actually obtain an analytic expression for $F(t, T_1, T_2)$. We calculate

(20) $\quad (T_2 - T_1) \times F(t, T_1, T_2)$

$$= \int_{T_1}^{T_2} \Lambda(u)\, du + \mathbf{e}_1^{\mathrm{T}} \int_{T_1}^{T_2} e^{A(u-t)}\, du \mathbf{X}(t)$$

$$- i\psi_L'(-i\theta) \mathbf{e}_1^{\mathrm{T}} A^{-1} \int_{T_1}^{T_2} (e^{A(u-t)} - I)\, du \mathbf{e}_p$$

$$= \int_{T_1}^{T_2} \Lambda(u)\, du + \mathbf{e}_1^{\mathrm{T}} A^{-1} \left(e^{A(T_2-t)} - e^{A(T_1-t)} \right) \mathbf{X}(t)$$

$$- i\psi_L'(-i\theta) \mathbf{e}_1^{\mathrm{T}} A^{-1} \left(A^{-1} e^{A(T_2-t)} - A^{-1} e^{A(T_1-t)} - (T_2 - T_1)I \right) \mathbf{e}_p .$$

A similar analytic expression in the geometric case seems hard to obtain, if possible.

4. The Spot-forward Relationship

The main objective in this section is to represent the forward price explicitly in terms of the spot price at time t. As we shall see, this will involve the derivatives of the spot price dynamics, and the forward price will become a linear combination of specific term structures scaled by derivatives of the spot.

To this end, denote by $X_i(t) = \mathbf{e}_i^{\mathrm{T}} \mathbf{X}(t)$, for $i = 1, \ldots, p$. Obviously, $X_i(t)$ will be the ith coordinate of $\mathbf{X}(t)$, and in particular we have that $X_1(t) = \mathbf{e}_1^{\mathrm{T}} \mathbf{X}(t) = Y(t)$. Consider the term $\mathbf{e}_1^{\mathrm{T}} e^{A(T-t)} \mathbf{X}(t)$ in the forward price dynamics in Propositions. 3.1–3.2. We find

$$\mathbf{e}_1^{\mathrm{T}} e^{A(T-t)} \mathbf{X}(t) = \sum_{i=1}^{p} f_i(T - t) X_i(t),$$

where

(21) $$f_i(x) = \mathbf{e}_1^{\mathrm{T}} e^{Ax} \mathbf{e}_i ,$$

for $i = 1, \ldots, p$. Observe that $f_i(0) = \mathbf{e}_1^{\mathrm{T}} \mathbf{e}_i$ which is zero for $i > 1$ and one otherwise. Now, recall from Proposition 2.1 that the kth derivative of $Y(t)$ exists for $k = 1, \ldots, p - 1$, and that

$$Y^{(k)}(t) = \int_{-\infty}^{t} \mathbf{e}_{k+1}^{\mathrm{T}} e^{A(t-s)} \mathbf{e}_p\, dL(s) = \mathbf{e}_{k+1}^{\mathrm{T}} \mathbf{X}(t) = X_{k+1}(t) .$$

Hence, we have shown the following Proposition:

Proposition 4.1. *Let $f_i(x)$ be defined as in (21) for $i = 1, \ldots, p$. If S is an arithmetic spot price as in (12), then*

$$f(t, T) = \Lambda(T) + \sum_{i=1}^{p} f_i(T - t) Y^{(i-1)}(t) - i\psi_L'(-i\theta) \mathbf{e}_1^{\mathrm{T}} A^{-1} \left(e^{A(T-t)} - I \right) \mathbf{e}_p ,$$

for $0 \le t \le T$. If the spot price is a geometric model as in (13), then

$$\ln f(t, T) = \Lambda(T) + \sum_{i=1}^{p} f_i(T - t)Y^{(i-1)}(t) + \int_0^{T-t} \psi_{L,\theta}\left(-i\mathbf{e}_1^\top e^{As}\mathbf{e}_p\right) ds,$$

for $0 \le t \le T$.

Consider the arithmetic spot model case: We see that the forward price will depend explicitly on the deseasonalized spot price and its derivatives up to order $p - 1$. If we have that the seasonal function Λ is sufficiently differentiable, we can rewrite this into a dependency on the spot price and its derivatives up to order $p - 1$. This result is very different from the classical spot-forward relationship, where the forward price is simply proportional to the current spot price only. In our setting we find that also the rate of growth, the curvature etc. of the spot price matter in the forward price. This is a result of our choice of spot price model being a CAR(p) dynamics, along with the non-tradeability assumption of the spot.

We next investigate the term structure shapes defined by the functions $f_i(x)$ in (21). As a case study, we take parameters from the fitting of the CAR(3)-model to daily average temperature data collected over more than 40 years in Stockholm, Sweden. The statistical estimation procedure along with the estimates are all reported in Benth et al. [9], and of particular interest here is the parameters in the CAR(3)-model. It was found that the α-parameters of the A matrix become

$$\alpha_1 = 2.043, \quad \alpha_2 = 1.339, \quad \alpha_3 = 0.177.$$

These values give the eigenvalues $\lambda_1 = -0.175$ and $\lambda_{2,3} = -0.934 \pm 0.374i$ for A, yielding a stationary CAR(3)-model. In Figure 4 we have plotted the resulting f_1, f_2 and f_3 defined in (21) as a function of time to maturity x.

All three functions f_1, f_2 and f_3 tend to zero as time to maturity goes to infinity, so in the long end of the forward market the contribution from these functions will become negligible. In the short end, that is, for small times to maturity x, the main contribution comes from f_1, as the two others start at zero for $x = 0$. We also clearly see that f_2 is bigger than f_1 and f_3 around its peak at $x \approx 3$.

We may view the functions f_i as *template* forward curves, which give the shape scaled by the corresponding values of $\mathbf{X}(t)$, that is, the value and its derivatives of the (deseasonalized) temperature. Thinking in terms of principal component analysis, we have that f_1 gives the *level* of the forward curve, corresponding to a shape decreasing from one towards zero in an exponential fashion. The template curve f_2 is scaled by the derivative of the (deseasonalized) temperature, and hence one can interpret f_2 as the *slope* in this context. The curve f_2 is increasing towards a maximum value, after which it decreases to zero in a seemingly exponential way. It contributes with a hump in the overall forward curve. If the temperature is increasing (has a positive) slope, there will be an upward pointing hump in the

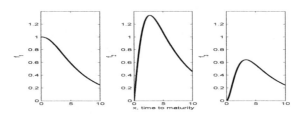

Figure 4. f_i for $i = 1, 2, 3$ for Stockholm, Sweden.

curve, while a temperature experiencing a decline at time t will yield a downward pointing hump. This hump will be most significant at around 3 days to maturity, and obviously the size is determined by how strong the slope of the temperature is at the time in question. The final template f_3 is scaled by the double-derivative of the (deseasonalized) temperature, and thus we relate this to the *curvature*. The curve f_3 will also contribute with a hump, which will point upward in the case of a convex temperature, and downward pointing if the temperature is concave. We see that the hump is smaller than for f_2, and f_2 is the template that contributes most among the three when x is around 3. This means that the slope of the temperature is more important than level and curvature for times to maturity at around 3. Also, we see that the shape of f_3 is different in the very short end, with a significantly smaller increase than f_2. In fact, f_2 is concave for small x, while f_3 in convex. Note that an increasing but concave temperature at time t (positive derivative, but negative double-derivative) will dampen the hump in the overall forward curve, while an increasing temperature being convex yields possibly a large hump in the short end of the curve. In Figure 5 we have plotted these two cases for some illustrative values of the vector $\mathbf{X}(t)$. Note that in this plot we have ignored the contribution from the seasonality function and other terms, and only focused on the part given by the templates f_1, f_2 and f_3.

We have chosen $Y(t) = 3$, that is, the current temperature is three degrees above its mean. The slope of the temperature is ± 1.5, meaning that the temperature is rapidly increasing or decreasing. Finally, we have used $Y''(t) = \pm 0.5$. We observe that for a negative curvature (concave temperature), we have decreasing forward curves, with the one having negative slope being smallest. If the temperature is convex (positive curvature), we get a hump in the forward curve, again the smallest curve stemming from a negative slope in the temperature.

To gain further insight into the shape of $f_i(x)$, $i = 1, \ldots, p$ defined in (21), we apply the spectral representation of A to re-express it into a sum of exponentials. To do this, let us first assume that A has p distinct eigenvalues $\lambda_1, \ldots, \lambda_p$, with corresponding eigenvectors $\mathbf{v}_1, \ldots, \mathbf{v}_p$. One easily verifies that

$$\mathbf{v}_j = (1, \lambda_j, \lambda_j^2, \ldots, \lambda_j^{p-1})^{\mathrm{T}},$$

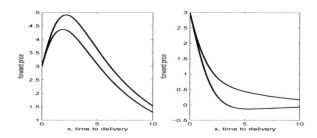

Figure 5. Forward curves as a function of time to delivery for various combinations of $Y'(t)$ and $Y''(t)$ with $Y(t) = 3$. To the left, we have $Y'(t) = \pm 1.5$ and $Y''(t) = 0.5$, whereas to the right we have $Y'(t) = \pm 1.5$ and $Y''(t) = -0.5$. The highest value of $Y'(t)$ gives the largest values of the forward price.

for $j = 1, \ldots, p$. Letting the matrix C consist of columns being the eigenvectors, we find that $\mathbf{e}_i = \sum_{j=1}^{p} a_j^i \mathbf{v}_j$ where the vector $\mathbf{a}_i = (a_1^i, \ldots, a_p^i)^{\mathrm{T}}$ is given as $\mathbf{a}_i = C^{-1}\mathbf{e}_i$. Therefore,

$$\mathbf{e}_i = \sum_{j=1}^{p} (\mathbf{e}_j^{\mathrm{T}} C^{-1}\mathbf{e}_i)\mathbf{v}_j .$$

Hence, we obtain

$$f_i(x) = \mathbf{e}_1^{\mathrm{T}} e^{Ax} \mathbf{e}_i = \sum_{j=1}^{p} (\mathbf{e}_j^{\mathrm{T}} C^{-1}\mathbf{e}_i) e^{\lambda_j x}$$

for a given $i = 1, \ldots, p$. We see that the shapes of all $f_i(x)$ can be represented as a weighted sum of exponentials. Due to stationarity, these exponentials are decaying as functions of time to maturity, at the speed determined by the real part of the eigenvalues. Each exponential term is scaled by factors $\mathbf{e}_j^{\mathrm{T}} C^{-1}\mathbf{e}_i$, or, the jith element of the inverse of the eigenvectors matrix C.

Let us consider forwards with delivery period, like forwards on the average temperature or electricity spot price over a given period. Letting S be defined as the arithmetic spot price model defined in (12), we recall the expression for the forward price $F(t, T_1, T_2)$ in (20), where the averaging takes place in the time interval $[T_1, T_2]$. We find that

$$(22) \qquad F(t, T_1, T_2) = \frac{1}{T_2 - T_1} \int_{T_1}^{T_2} \Lambda(u)\, du + \sum_{i=1}^{p} F_i(T_1 - t, T_2 - T_1) Y^{(i-1)}(t)$$
$$+ \Psi(T_1 - t, T_2 - T_1),$$

where

$$(23) \qquad \Psi(x, y) = -\mathrm{i}\psi_L'(-\mathrm{i}\theta)(\mathbf{e}_1^{\mathrm{T}} A^{-2} \frac{1}{y}\left(e^{Ay} - I\right) e^{Ax}\mathbf{e}_p - \mathbf{e}_1^{\mathrm{T}} A^{-1}\mathbf{e}_p),$$

and

$$
(24) \qquad F_i(x, y) = \mathbf{e}_1^{\mathrm{T}} A^{-1} \frac{1}{y} \left(e^{Ay} - I \right) e^{Ax} \mathbf{e}_i \,,
$$

for $i = 1, \ldots, p$. Here, $x = T_1 - t$ denotes *time to delivery period starts* and $y = T_2 - T_1$ *length of delivery period*. Note that $F_i(0, y) \neq 0$ for all $i = 1, \ldots, p$. This is a reflection that $F(t, T_1, T_2)$ is not converging to the underlying spot as $t \to T_1$, due to the the delivery period. Such a behaviour is particular in forward markets with delivery period, contrary to "classical" commodity markets where the forward is equal to the spot when time to delivery is zero. In Figure 6 we have plotted the templates F_1, F_2 and F_3 for the case of monthly "delivery" period, that is, a forward on the average over a month on the underlying. This would correspond to a monthly temperature forward as the A matrix is also here borrowed from Stockholm, and we see that all three curves are decaying. Interestingly, the dominating factor will be F_2, which is scaled by the derivative of the deseasonalized temperature. Hence, the forward curve is most sensitive to the *change* in temperature, and not the level or the curvature. We also see no humps.

In Figure 7 we choose y to be one week, and see that there are humps coming into the forward curve stemming from the *slope* and *curvature*. Again the slope is significantly more important in contributing to the forward curve than the other two values.

Note that at the Chicago Mercantile Exchange, there is trade in temperature forwards settled over one week.

Next, let us consider how the forward price dynamics is evolving as a function of the spot price empirically. We consider a numerical example, simulating the spot price path by the path of $\mathbf{X}(t)$, which can be done exact as this is an Ornstein-Uhlenbeck process. Hence, we can simulate the path of a forward price $t \mapsto f(t, T)$ for a given delivery time T. To this end, from (6) we find, for $\Delta > 0$,

$$
(25) \qquad \mathbf{X}(t + \Delta) = e^{A\Delta}\mathbf{X}(t) + \int_t^{t+\Delta} e^A(t + \Delta - s)\mathbf{e}_p \, dL(s) \,.
$$

Hence, we can simulate $\mathbf{X}(t + \Delta)$ from $\mathbf{X}(t)$ and an independent noise given by the stochastic integral $\int_t^{t+\Delta} \exp(A(t + \Delta - s))\mathbf{e}_p \, dL(s)$. In the case $L = \sigma B$, a Brownian motion with volatility σ, this stochastic integral is a p-dimensional Gaussian random variable with mean zero and variance given by the Itô isometry as

$$
\mathrm{Var}\!\left(\int_t^{t+\Delta} \sigma \exp(A(t + \Delta - s))\mathbf{e}_p \, dB(s) \right)
$$
$$
= \sigma^2 \int_t^{t+\Delta} e^{A(t+\Delta-s)}\mathbf{e}_p\mathbf{e}_p^{\mathrm{T}}e^{A^{\mathrm{T}}(t+\Delta-s)} \, ds
$$
$$
= \sigma^2 \int_0^{\Delta} e^{As}\mathbf{e}_p\mathbf{e}_p^{\mathrm{T}}e^{A^{\mathrm{T}}s} \, ds \,.
$$

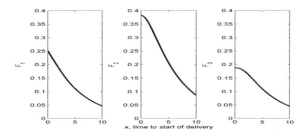

Figure 6. F_i for $i = 1, 2, 3$ for Stockholm, Sweden with y being set to one month.

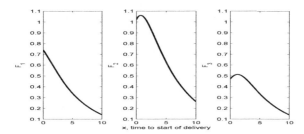

Figure 7. F_i for $i = 1, 2, 3$ for Stockholm, Sweden with y being set to one week.

Hence, the noises are independent and identically distributed, being only a function of the time step Δ.

We simulated the dynamics of $\mathbf{X}(t)$ for $p = 3$ with the matrix A as before and with $L = \sigma B$ where $\sigma = 1$. The time step was chosen to be $\Delta = 0.1$, measured in days, and in Figure 8 we have plotted the path of $Y(t)$ over the time interval 0 to 30 days, along with the corresponding forward price for a contract with delivery $T = 30$. We have assumed the seasonality Λ being identically equal to zero and supposed zero market price of risk $\theta = 0$. As it is evident from the plot, far from maturity there is essentially no variation in the forward price, a result of the stationarity of Y (or, more precisely, of $\mathbf{X}$). Closer to maturity, the variations in the forward price become bigger, and we see how they follow the slope and level of the spot price $Y(t)$. It is harder to see the effect of the curvature directly. But interestingly, it seems that the forward price path is much rougher than that of the spot. This can be attributed to the fact that the spot is twice differentiable, whereas the forward is explicitly depending on all the coordinates of $\mathbf{X}(t)$, in particular $X_3(t) = \mathbf{e}_3^T\mathbf{X}(t)$ which is not differentiable.

A more realistic situation is when we only observe the path of the spot, and we must recover its derivatives in order to compute the forward price. As the

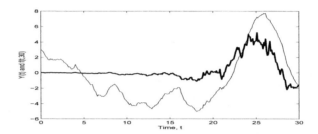

Figure 8. The path of $Y(t)$ along with $f(t, T)$ where $T = 30$. Seasonality and market price of risk are assumed to be zero. The spot path $Y(t)$ is the thin curve.

dynamics of the spot $Y(t)$ is on a state-space form, we could use a Kalman filter for this purpose (see Benth and Šaltytė Benth [8]). However, we can also use numerical differentiation of the past and present spot observations. Backward finite differencing yields the approximations

(26)
$$Y'(t) \approx \frac{Y(t) - Y(t - \Delta)}{\Delta}$$

(27)
$$Y''(t) \approx \frac{Y(t) - 2Y(t - \Delta) + Y(t - 2\Delta)}{\Delta^2} .$$

We applied this routine on a simulated example to check its performance. In Figure 9 we have simulated the path of $Y(t)$ for the same set of parameters as above and applied the finite differences to recover the first and second derivative (depicted as dotted lines in the figure). To be more in line with applications, we assume that we have daily observations of Y, and simulated a path over 100 days. In the figure, we have included the actual paths of $X_2(t) = \mathbf{e}_2^T\mathbf{X}(t)$ and $X_3(t) = \mathbf{e}_3^T\mathbf{X}(t)$ realized from the simulation (depicted as complete lines on the figure). We see that the finite difference approximations of the path of $Y(t)$ are recovering the actual derivatives very well, motivating that this procedure would make sense in practical applications.

5. Concluding Remarks

In this paper we have consider the forward price dynamics as a function of the spot price modelled as a continuous-time autoregressive dynamics. In the case of autoregression of order $p > 1$, the forward curve is driven by the the derivatives of the spot price up to order $p - 1$, where each component gives a contribution to the structure of the forward curve. The components may be viewed as templates for the forward curve, as they are the basis functions for the possible shapes that can be achieved by this model. The forward curve can vary between contango and backwardation, as well as incorporating humps.

The different forward term structures appear depending on the state of the spot and its derivatives. In fact, we may explain humps in the forward curve

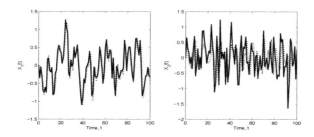

Figure 9. Estimation of $Y'(t)$ (left) and $Y''(t)$ (right) based on finite differencing of the path of $Y(t)$ (dotted lines), versus the exact first and second derivatives (complete lines).

as appearing stochastically as a result of the underlying spot having a positive slope (that is, the spot price is currently increasing). In many "physical markets" like weather and energy/commodities, one can identify trends in spot prices (by technical analysis, say), and this may be applied to identify and explain occurence of humps. It would be interesting to see whether this holds true in an empirical setting. After all, stochastic volatility may explain humps in the forward curve as well (see Benth [3]), and this raises the question whether a hump occured because of the state of the volatility, or due to some strucutral propetrties of the spot price path (or both).

We also would like to refer to a study of Diebold and Li [12] which uses the popular Nelson-Siegel yield curve in the context of forward rates and drives the components of this curve by autoregressive time series of order 1. Our study gives results on the forward process in the same spirit as this study, however, in a specific framework that ensures an arbitrage-free forward price dynamics.

In a future study we will extend our analysis to general Lévy semistationary dynamics of the spot, which would allow for a moving average structure in the continuous-time autoregressive model. It is expected that in this case that the forward dynamics will depend on the history of the spot, and not only on the current value and its derivatives.

Acknowledgement

The authors acknowledge financial support from the project "Energy Markets: Modelling, Optimization and Simulation (EMMOS)" funded by the evita program of the Norwegian Research Council.

References

1. Barndorff, O. E. Benth, F. E. and Veraart, A. E. D. (2010), Ambit Processes and Stochastic Partial Differential Equations. In: Di Nunno, G. and Øksendal, B. (eds.), *Advanced mathematical methods for finance*, Springer-Verlag Berlin Heidelberg 2011, 35–73.

2. Basse, A. and Pedersen, J. (2009), "Lévy driven moving averages and semimartingales," *Stoch. Proc. Appl.*, **119**, 2970–2991.

3. Benth, F. E. (2011), "The stochastic volatility model of Barndorff-Nielsen and Shephard in commodity markets," *Math. Finance*, **21**(4), 595–625.

4. Benth, F. E. and Eyjolfsson, H. (2013), "Stochastic modeling of power markets using stationary processes," To appear in the *Ascona Proccedings 2011*. *Progress in Probability*, R. Dalang, M. Dozzi and F. Russo (eds.), Birkhäuser Verlag.

5. Benth, F. E. Klüppelberg, Müller, G. and Vos, L. (2012), "Futures pricing in electricity markets based on stable CARMA spot models," Available on arxive at `http://arxiv.org/abs/1201.1151`.

6. Benth, F. E. Koekebakker, S. and Taib, I. B. C. M. (2012), "Stochastic modelling of spot freight rates," Submitted manuscript.

7. Benth, F. E. and Šaltytė Benth, J. (2009), "Dynamic pricing of wind futures," *Energy Econ.*, **31**(1), 16–24.

8. Benth, F. E. and Šaltytė Benth, J. (2013), *Modeling and Pricing in Financial Markets for Weather Derivatives*, World Scientific, Singapore.

9. Benth, F. E. Šaltytė Benth, J. and Koekebakker, S. (2008), *Stochastic Modelling of Electricity and Related Markets*, World Scientific, Singapore.

10. Bingham, N. H. and Kiesel, R. (2010), *Risk-Neutral Valuation: Pricing and Hedging of FInancial Derivatives*, Second ed., Springer Verlag.

11. Brockwell, P. J. (2001), "Lévy driven CARMA processes," *Ann. Inst. Statist. Math.*, **53**(1), 113–124.

12. Diebold, F. X., and Li, C. (2006), "Forecasting the term structure of government bond yields," *J. Econometrics*, **130**, 337–364.

13. Doob, J. L. (1944), "The elementary Gaussian processes," *Ann. Math. Statist.*, **15**(3), 229–282.

14. Folland, G. B. (1984), *Real Analysis*, Wiley, Chichester.

15. Garcia, I. Klüppelberg, C. and Müller, G. (2010), "Estimation of stable CARMA models with an application to electricity spot prices," *Statistical Modelling*, **11**(5), 447–470.

16. Härdle, W. and Lopez-Cabrera, B. (2012), "The implied market price of weather risk," *Appl. Math. Finance*, **19**(1), 59–95.

17. Ikeda, N. and Watanabe, S. (1981), *Stochastic Differential Equations and Diffusion Processes*, North-Holland/Kodansha.

18. Karatzas, I. and Shreve, S. E. (1991), *Brownian Motion and Stochastic Calculus*, Second Edition, Springer Verlag, New York.

19. Paschke, R. and Prokopzcuk, M. (2010), "Commodity derivatives valuation with autoregressive and moving average components in the price dynamics," *J. Banking & Finance*, **34**(11), 2741–2752.

20. Protter, Ph. (1990), *Stochastic Integration and Differential Equations*, Springer Verlag, New York.

21. Sato, K. -I. (1999), *Lévy Processes and Infinitely Divisible Distributions*, Cambridge University Press.

22. Zakamouline, V. Benth, F. E. and Koekebakker, S. (2010), "A continuous time model for interest rate with autoregressive and moving average components," *AIP Conference Proceedings*, September 30, 2010, Volume **1281**, 531–534.

A Bottom-Up Dynamic Model of Portfolio Credit Risk.
Part I: Markov Copula Perspective

Tomasz R. Bielecki[1]*, Areski Cousin[2]†, Stéphane Crépey[3]‡
Alexander Herbertsson[4]§

[1]Department of Applied Mathematics, Illinois Institute of Technology,
Chicago, IL 60616, USA
[2] Université de Lyon, Université Lyon 1, LSAF, France
[3] Laboratoire Analyse et Probabilités, Université d'Évry Val d'Essonne,
91037 Évry Cedex, France
[4] Centre for finance/Department of Economics, University of Gothenburg,
SE 405 30 Göteborg, Sweden

We consider a bottom-up Markovian copula model of portfolio credit risk where instantaneous contagion is possible in the form of simultaneous defaults. Due to the Markovian copula nature of the model, calibration of marginals and dependence parameters can be performed separately using a two-steps procedure, much like in a standard static copula set-up. In this sense this model solves the bottom-up top-down puzzle which the CDO industry had been trying to do for a long time. It can be applied to any dynamic credit issue like consistent valuation and hedging of CDSs, CDOs and counterparty risk on credit portfolios.

Key words: Portfolio credit risk, Credit derivatives, Markov copula model, Common shocks, Dynamic hedging.

*The research of T.R. Bielecki was supported by NSF Grant DMS–0604789 and NSF Grant DMS–0908099.

†The research of A. Cousin benefited from the support of the DGE and the ANR project Ast&Risk.

‡The research of Stéphane Crépey benefited from the support of the "Chair Markets in Transition" under the aegis of Louis Bachelier laboratory, a joint initiative of École polytechnique, Université d'Évry Val d'Essonne and Fédération Bancaire Française.

§The research of A. Herbertsson was supported by the Jan Wallander and Tom Hedelius Foundation and by Vinnova.

1. Introduction

The CDO market has been deeply and adversely impacted by the crisis. In particular, CDO issuances have become quite rare. Nevertheless, there are huge notionals of CDO contracts outstanding and market participants continue to be confronted with the task to hedge their positions in these contracts up to maturity date. Moreover, according to the regulation (see [2]), tranches on standard indices and their associated liquid hedging positions continue to be charged as hedgesets under internal VaR-based method. Regarding the CDO hedging issue we refer the reader to Laurent, Cousin and Fermanian [28], Frey and Backhaus [23], Cont and Kan [15] or Cousin, Crépey and Kan [17]. In particular it has been established empirically in [15] and [17] that a single-instrument hedge of a CDO tranche by the corresponding credit index is often not good enough. In this and the companion paper [5], we deal with a bottom-up Markovian copula model, in which hedging loss derivatives by single-name instruments can be performed in a theoretically sound and practical way.

There are two major theoretical contributions of these papers:

- In this paper, we construct a Markov model where dependence between default risks derives from the possibility of joint defaults. The Markovian structure of the model is adequate for the problem at hand, that is for the problem of dynamic hedging of portfolio credit risk. The (dynamic) copula property of the model allows for separation of calibration of the univariate marginals of the underlying multivariate Markov process, from calibration of the dependence structure between the components of the process. This is of critical importance from the practical point of view.

- We show that the conditional dependence structure of default times belongs to the class of Marshall-Olkin copulas (see Prop 2.5). This result is exploited in [5] to construct an equivalent conditional factor representation of our Markovian model which relies on "common shocks", the latter being represented by Cox processes likely to trigger defaults simultaneously in some pre-specifed group of obligors. This is important from the practical point of view as this interpretation underlies semi-explicit convolution-based pricing schemes to assess the credit portfolio loss distribution at several time horizons. Such numerical schemes play a crucial role when calibrating credit portfolio models and in related applications such as hedging portfolio credit derivatives by individual names, or counterparty risk valuation for portfolios (see [1, 7]).

The common shock aspect of our model is related to the work by Elouerkhaoui [21] (see also Brigo et al. [12, 13, 14]). Consequently, some results derived in this and the companion paper are consistent with results derived in [21]. However, there are major differences between our study and the one presented in [21]:

- Firstly, the approach of [21] suffers from the "curse of dimensionality" due to the need of summation (integration) over the set denoted by Π_n in [21] (see for example equation (2.6) therein, and compare with our own result (17) below), the set of all subsets of the set $\{1, 2, \ldots, n\}$. By contrast, the complexity of our formula (7) for the generator of our Markov process, or of our common shock algorithms described in Subsection 2.1 of the companion paper [5], are controlled by the cardinality of our shocks set $\mathcal{Y}$, typically a few units in applications (see [5]).

- Secondly, as already stated, our methodology allows for separation of calibration of idiosyncratic (marginal) laws of the underlying Markov process, from the calibration of the dependence structure of the process. The calibration really amounts to calibrating the infinitesimal generator of the underlying Markov process, and once this is done, the model can be used for consistent pricing and hedging of both the underlying products, such as CDO tranches, as well as options on such with future expiration dates (e.g. in the context of CVA computations); this feature obviously contributes to increased practical use of our methodology. In this sense, our Markov copula model is a genuine dynamic model, as a model of dependence between underlying stochastic processes. This is not really the case with the model developed in [21], where the "dynamic copula" feature is in the sense of Patton's conditional copula [31], which is a stochastic process itself, and as such can't be calibrated to initial data.

- Last, but not least, the Markov copula approach of this paper is generic in the sense that, as demonstrated in [10, 11], it also applies to modeling of dynamics of credit ratings. This is not the case with the approach of [21].

Comparing now our methodology to what is done in Brigo et al. [12, 13, 14], we see that the major differences can be summarized as follows:

- Our approach is a bottom-up approach, hence an approach applicable for hedging basket products using individual names, whereas the approach taken in [12, 13] is a top-down approach, and, as such, is not applicable for hedging basket products using individual names;

  – This also applies to the so-called GPCL extension of the model of [14] in which individual names are represented so that, in principle, hedging basket products using individual names could be considered in this setup. This is not practical however because fault of a suitable decoupling property between the dependence structure and the individual names in the model, the calibration of the model can only be addressed through a global joint optimization procedures involving all the model parameters at the same time, which is untractable numerically.

- Again, our approach is generic in the sense that it also applies to modeling of dynamics of credit ratings. This is not the case with the approach of [12, 13, 14].

This paper is organized as follows. In Section 2 we formulate a bottom-up Markovian copula model, in which individual default processes for various credit names are coupled together by means of simultaneous defaults. In Section 3 we exploit the dynamic structure of the model to derive explicit dynamic min-variance hedging formulas. The more technical proofs are deferred to Appendix A. The algorithmic aspects of the model based on the common-shock representation, as well as illustrative numerics, are provided in the companion paper [5]. Fine features of the modeling of the default marginals (single-name modeling in different kinds of affine setups) are considered in [6]. A short announcing version of these results can be found in [4].

In the rest of the paper we consider a risk neutral pricing model $(\Omega, \mathcal{F}, \mathbb{P})$, for a filtration $\mathcal{F} = (\mathcal{F}_t)_{t\in[0,T]}$ which will be specified below and where $T \geq 0$ is a fixed time horizon. We denote $N_n = \{1, \ldots, n\}$ and let $\mathcal{N}_n$ denote the set of all subsets of N_n where n represents the number of obligors in the underlying credit portfolio. Further, we set $\max \emptyset = -\infty$.

2. Model of Default Times

In this section we construct a bottom-up Markovian model consisting of a multivariate factor process $\mathbf{X}$ and a vector $\mathbf{H}$ representing the default indicator processes in a pool of n different credit names. More specifically, $\mathbf{H}_t$ is a vector in $\{0, 1\}^n$ where the i-th entry of $\mathbf{H}_t$ is the indicator function for the event of a default of obligor i up to time t. The purpose of the factor process $\mathbf{X}$ is to more realistically model diffusive randomness of credit spreads.

In our model, defaults are the consequence of some "shocks" associated with groups of obligors. We define the following pre-specified set of groups

$$\mathcal{Y} = \{\{1\}, \ldots, \{n\}, I_1, \ldots, I_m\},$$

where $I_1, \ldots, I_m$ are subsets of $\{1, \ldots, n\}$, and each group I_j contains at least two obligors or more. The shocks are divided in two categories: the "idiosyncratic" shocks associated with singletons $\{1\}, \ldots, \{n\}$ can only trigger the default of name $1, \ldots, n$ individually, while the "systemic" shocks associated with multi-name groups $I_1, \ldots, I_m$ may simultaneously trigger the default of all names in these groups. Note that several groups I_j may contain a given name i, so that only the shock occurring first effectively triggers the default of that name. As a result, when a shock associated with a specific group occurs at time t, it only triggers the default of names that are still alive in that group at time t. In the following, the elements Y of $\mathcal{Y}$ will be used to designate shocks and we let $\mathcal{I} = (I_l)_{1 \leq l \leq m}$ denote the pre-specified set of multi-name groups of obligors.

Let $\nu = |\mathcal{Y}| = n+m$ denote the cardinality of $\mathcal{Y}$. Given a multivariate Brownian motion $\mathbf{W} = (W^Y)_{Y \in \mathcal{Y}}$ with independent components, we assume that the factor process $\mathbf{X} = (X^Y)_{Y \in \mathcal{Y}}$ is a strong solution to

$$(1) \qquad dX_t^Y = b_Y(t, X_t^Y)\, dt + \sigma_Y(t, X_t^Y)\, dW_t^Y,$$

for suitable drift and diffusion functions $b_Y = b_Y(t, x)$ and $\sigma_Y = \sigma_Y(t, x)$. By application of Theorem 32 page 100 of Protter [32], this makes $\mathbf{X}$ an $\mathcal{F}^{\mathbf{W}}$-Markov process admitting the following generator acting on functions $v = v(t, \mathbf{x})$ with $\mathbf{x} = (x_Y)_{Y \in \mathcal{Y}}$

$$(2) \qquad A_t v(t, \mathbf{x}) = \sum_{Y \in \mathcal{Y}} \left(b_Y(t, x_Y) \partial_{x_Y} v(t, \mathbf{x}) + \tfrac{1}{2} \sigma_Y^2(t, x_Y) \partial_{x_Y^2}^2 v(t, \mathbf{x}) \right).$$

Let $\mathcal{F} := \mathcal{F}^{(\mathbf{W}, \mathbf{H})}$ be the filtration generated by the Brownian motion $\mathbf{W}$ and the point process $\mathbf{H}$. Given the "intensity functions" of shocks, say $\lambda_Y = \lambda_Y(t, x_Y)$ for every shock $Y \in \mathcal{Y}$, we would like to construct a model in which the $\mathcal{F}$-predictable intensity of a jump of $\mathbf{H} = (H^i)_{1 \leq i \leq n}$ from $\mathbf{H}_{t-} = \mathbf{k}$ to $\mathbf{H}_t = \mathbf{l}$, with $\mathbf{l} \neq \mathbf{k}$ in $\{0, 1\}^n$, is given by

$$(3) \qquad \lambda(t, \mathbf{X}_t, \mathbf{k}, \mathbf{l}) := \sum_{\{Y \in \mathcal{Y};\, \mathbf{k}^Y = \mathbf{l}\}} \lambda_Y(t, X_t^Y),$$

where, for any $Z \in \mathcal{N}_n$, the expression $\mathbf{k}^Z$ denotes the vector obtained from $\mathbf{k} = (k_1, \ldots, k_n)$ by replacing the components k_i, $i \in Z$, by numbers one (whenever k_i is not equal to one already). The intensity of a jump of $\mathbf{H}$ from $\mathbf{k}$ to $\mathbf{l}$ at time t is thus equal to the sum of the intensities of the shocks $Y \in \mathcal{Y}$ such that, if the joint default of the survivors in group Y occurred at time t, then the state of $\mathbf{H}$ would move from $\mathbf{k}$ to $\mathbf{l}$.

Example 2.1. Figure 1 shows one possible defaults path in our model with $n = 5$ and $\mathcal{Y} = \{\{1\}, \{2\}, \{3\}, \{4\}, \{5\}, \{4, 5\}, \{2, 3, 4\}, \{1, 2\}\}$. The inner oval shows which common-shock happened and caused the observed default scenarios at successive default times. At the first instant, default of name 2 is observed as the consequence of the idiosyncratic shock $\{2\}$. At the second instant, names 4 and 5 have defaulted simultaneously as a consequence of the systemic shock $\{4, 5\}$. At the fourth instant, the systemic shock $\{2, 3, 4\}$ triggers the default of name 3 alone as name 2 and 4 have already defaulted. At the fifth instant, default of name 1 alone is observed as the consequence of the systemic shock $\{1, 2\}$. Note that the information produced by the arrival of the shock-events cannot be deduced from the mere observation of the sequence of states followed by $\mathbf{H}_t$.

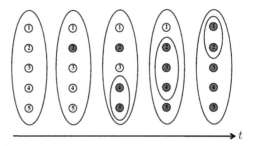

Figure 1. One possible defaults path in a model with $n = 5$ and $\mathcal{Y} = \{\{1\},\{2\},\{3\},\{4\},\{5\},\{4,5\},\{2,3,4\},\{1,2\}\}$.

To achieve (3) we follow the classical methodology: we construct **H** by an **X**-related change of probability measure, starting from a continuous-time Markov chain with intensity one. This construction is detailed in Appendix A.1.

2.1   Itô Formula

In this subsection we state the Itô formula for functions of the Markov process $(\mathbf{X}, \mathbf{H})$.

For any set $Z \in \mathcal{N}_n$, let the set-event indicator process H^Z denote the indicator process of a joint default of the names in Z and only in Z. For $\mathbf{k} = (k_1, \ldots, k_n) \in \{0, 1\}^n$, we introduce $\text{supp}(\mathbf{k}) = \{i \in \mathcal{N}_n; \, k_i = 1\}$ and $\text{supp}^c(\mathbf{k}) = \{i \in \mathcal{N}_n; \, k_i = 0\}$. Hence, $\text{supp}(\mathbf{k})$ denotes the obligors who have defaulted in state $\mathbf{k}$ and similarly $\text{supp}^c(\mathbf{k})$ are the survived names in the portfolio-state $\mathbf{k}$.

The following lemma provides the structure of the so called compensated set-event martingales M^Z, which we will use later as fundamental martingales to represent the pure jump martingale components of the various price processes involved.

Lemma 2.1. *For every set $Z \in \mathcal{N}_n$ the intensity of H^Z is given by $\ell_Z(t, \mathbf{X}_t, \mathbf{H}_t)$, so*

$$dM_t^Z = dH_t^Z - \ell_Z(t, \mathbf{X}_t, \mathbf{H}_t)dt$$

is a martingale, and the set-event intensity function $\ell_Z(t, \mathbf{x}, \mathbf{k})$ is defined as

$$\text{(4)} \qquad \ell_Z(t, \mathbf{x}, \mathbf{k}) = \sum_{Y \in \mathcal{Y}; \, Y \cap supp^c(\mathbf{k})=Z} \lambda_Y(t, x_Y).$$

Proof. See Appendix A.1.1.

So $\ell_Z(t, \mathbf{X}_t, \mathbf{H}_{t-}) = \sum_{Y \in \mathcal{Y}; \, Y_t=Z} \lambda_Y(t, X_t^Y)$, where for every Y in $\mathcal{Y} = \{\{1\}, \ldots, \{n\}, I_1, \ldots, I_m\}$ we define

$$\text{(5)} \qquad Y_t = Y \cap \text{supp}^c(\mathbf{H}_{t-}),$$

the set-valued process representing the survived obligors in Y right before time t. Let also $\mathcal{Z}_t = \{Z \in N_n; Z = Y_t$ for at least one $Y \in \mathcal{Y}\} \setminus \emptyset$ denote the set of all non-empty sets of survivors of sets Y in $\mathcal{Y}$ right before time t.

We now derive a version of the Itô formula, which is relevant for our model. It will be used below for establishing the Markov properties of our set-up, as well as for deriving price dynamics. Let $\sigma(t, \mathbf{x})$ denote the diagonal matrix with diagonal $(\sigma_Y(t, x_Y))_{Y \in \mathcal{Y}}$. Given a function $u = u(t, \mathbf{x}, \mathbf{k})$ with $\mathbf{x} = (x_Y)_{Y \in \mathcal{Y}}$ and $\mathbf{k} = (k_i)_{1 \leq i \leq n}$ in $\{0, 1\}^n$, we denote $\nabla u(t, \mathbf{x}, \mathbf{k}) = (\partial_{x_Y} u(t, \mathbf{x}, \mathbf{k}), Y \in \mathcal{Y})$, the (row-)gradient of u with respect to $\mathbf{x}$. Let also δu^Z represent the sensitivity of u to the event $Z \in N_n$, so

$$\delta u^Z(t, \mathbf{x}, \mathbf{k}) = u(t, \mathbf{x}, \mathbf{k}^Z) - u(t, \mathbf{x}, \mathbf{k}).$$

Proposition 2.2. *Given a regular enough function $u = u(t, \mathbf{x}, \mathbf{k})$, one has*

(6)
$$du(t, \mathbf{X}_t, \mathbf{H}_t) = \left(\partial_t + \mathcal{A}_t\right)u(t, \mathbf{X}_t, \mathbf{H}_t)dt + \nabla u(t, \mathbf{X}_t, \mathbf{H}_t)\sigma(t, \mathbf{X}_t)d\mathbf{W}_t$$
$$+ \sum_{Z \in \mathcal{Z}_t} \delta u^Z(t, \mathbf{X}_t, \mathbf{H}_{t-})dM_t^Z,$$

where

(7)
$$\mathcal{A}_t u(t, \mathbf{x}, \mathbf{k}) = \sum_{Y \in \mathcal{Y}} \left(b_Y(t, x_Y)\partial_{x_Y} u(t, \mathbf{x}, \mathbf{k}) + \frac{1}{2}\sigma_Y^2(t, x_Y)\partial_{x_Y^2}^2 u(t, \mathbf{x}, \mathbf{k})\right)$$
$$+ \sum_{Y \in \mathcal{Y}} \lambda_Y(t, x_Y)\delta u^Y(t, \mathbf{x}, \mathbf{k}).$$

Proof. See Appendix A.1.2.

In the Itô formula (6), the jump term may involve any of the 2^n set-events martingales M^Z for $Z \in N_n$. This suggests that the martingale dimension[1] of the model is $v + 2^n$, where $v = n + m$ corresponds to the dimension of the Brownian motion $\mathbf{W}$ driving the factor process $\mathbf{X}$ and 2^n corresponds to the jump component $\mathbf{H}$. Yet by a reduction which is due to specific structure of the intensities in our set-up, the jump term of $\mathcal{A}_t$ in (7) is a sum over the set of shocks $\mathcal{Y}$, which has cardinality v.

Note that our model excludes direct contagion effects in which intensities of surviving names would be affected by past defaults, as opposed to the bottom-up

---

[1] Minimal number of fundamental martingales which can be used as integrators to represent all the martingales in the model, see Appendix A.1.

contagion models treated by e.g. [16, 24, 25, 28]. To provide some understanding in this regard, we give a simple illustrative example.

Example 2.2. Take $N_n = \{1, 2, 3\}$, so that the state space of **H** contains 8 elements:

$$\{(0, 0, 0), (1, 0, 0), (0, 1, 0), (0, 0, 1), (1, 1, 0), (1, 0, 1), (0, 1, 1), (1, 1, 1)\}.$$

Now, let $\mathcal{Y}$ be given as $\mathcal{Y} = \{\{1\}, \{2\}, \{3\}, \{1, 2\}, \{1, 2, 3\}\}$. This is an example of the nested structure of $\mathcal{I}$ with $I_1 = \{1, 2\} \subset I_2 = \{1, 2, 3\}$. Suppose for simplicity that λ_Y does not depend either on t or on **x** (dependence in $t, \mathbf{x}$ will be dealt with in Subsection 2.2). Then, the generator $\mathcal{A}$ of the chain **H** is given in matrix-form by

$$
(8) \qquad \mathcal{A} \equiv
\begin{bmatrix}
\cdot & \lambda_{\{1\}} & \lambda_{\{2\}} & \lambda_{\{3\}} & \lambda_{\{1,2\}} & 0 & 0 & \lambda_{\{1,2,3\}} \\
0 & \cdot & 0 & 0 & \lambda_{\{2\}} + \lambda_{\{1,2\}} & \lambda_{\{3\}} & 0 & \lambda_{\{1,2,3\}} \\
0 & 0 & \cdot & 0 & \lambda_{\{1\}} + \lambda_{\{1,2\}} & 0 & \lambda_{\{3\}} & \lambda_{\{1,2,3\}} \\
0 & 0 & 0 & \cdot & 0 & \lambda_{\{1\}} & \lambda_{\{2\}} & \lambda_{\{1,2,3\}} + \lambda_{\{1,2\}} \\
0 & 0 & 0 & 0 & \cdot & 0 & 0 & \lambda_{\{3\}} + \lambda_{\{1,2,3\}} \\
0 & 0 & 0 & 0 & 0 & \cdot & 0 & \lambda_{\{2\}} + \lambda_{\{1,2,3\}} + \lambda_{\{1,2\}} \\
0 & 0 & 0 & 0 & 0 & 0 & \cdot & \lambda_{\{1\}} + \lambda_{\{1,2,3\}} + \lambda_{\{1,2\}} \\
0 & 0 & 0 & 0 & 0 & 0 & 0 & 0
\end{bmatrix}
$$

where '$\cdot$' represents the sum of all other elements in the row multiplied with -1. Now, consider group $\{1, 2, 3\}$. Suppose, that at some point of time obligor 2 is defaulted, but obligors 1 and 3 are still alive, so that process **H** is in state $(0, 1, 0)$. In this case the two survivors in the group $\{1, 2, 3\}$ may default simultaneously with intensity $\lambda_{\{1,2,3\}}$. Of course, here $\lambda_{\{1,2,3\}}$ cannot be interpreted as intensity of all three defaulting simultaneously, as obligor 2 has already defaulted. In fact, the only state of the model in which $\lambda_{\{1,2,3\}}$ can be interpreted as the intensity of all three defaulting, is state $(0, 0, 0)$. Note that obligor 1 defaults with intensity $\lambda_{\{1\}} + \lambda_{\{1,2,3\}} + \lambda_{\{1,2\}}$ regardless of the state of the pool, as long company 1 is alive. Similarly, obligor 2 will default with intensity $\lambda_{\{2\}} + \lambda_{\{1,2,3\}} + \lambda_{\{1,2\}}$ regardless of the state of the pool, as long company 1 is alive. Also, obligors 1 and 2 will default together with intensity $\lambda_{\{1,2,3\}} + \lambda_{\{1,2\}}$ regardless of the state of the pool, as long as company 1 and 2 still are alive.

2.2 Markov Copula Properties

Below, for every obligor i, a real-valued marginal factor process X^i will be defined as a suitable function of the above multivariate factor process $\mathbf{X} = (X^Y)_{Y \in \mathcal{Y}}$. We shall then state conditions on the intensities under which the marginal pair (X^i, H^i) is a Markov process. Markovianity of the model marginals (X^i, H^i) is crucial at the stage of calibration of the model, so that these marginals can be calibrated independently.

Observe that in view of (3), the intensity of a jump of H^i from $H^i_{t-} = 0$ to 1 is given by, for $t \in [0, T]$,

$$\sum_{\{Y \in \mathcal{Y}; i \in Y\}} \lambda_Y(t, X^Y_t), \tag{9}$$

where the sum in this expression is taken over all pre-specified shocks that can affect name i. We define the marginal factor X^i as a linear functional φ_i of the multivariate factor process $\mathbf{X} = (X^Y)_{Y \in \mathcal{Y}}$ so that $X^i_t := \varphi_i(\mathbf{X}_t)$. In general the transition intensity (9) implies non-Markovianity of the marginal (X^i, H^i). Hence, in order to make the process (X^i, H^i) to be Markov, one needs to impose a more specified parametrization of (9) as well as conditions on the mapping φ_i. To be more specific:

Assumption 2.3. For every obligor i, there exists a linear form $\varphi_i(\mathbf{x})$ and a real-valued function $\lambda_i(t, x)$ such that for every $(t, \mathbf{x})$ with $\mathbf{x} = (x_Y)_{Y \in \mathcal{Y}}$

$$\sum_{\{Y \in \mathcal{Y}; i \in Y\}} \lambda_Y(t, x_Y) = \lambda_i(t, \varphi_i(\mathbf{x})), \tag{10}$$

where, in addition, $X^i_t := \varphi_i(\mathbf{X}_t)$ is a Markov-process with respect to the filtration $\mathcal{F} = \mathcal{F}^{(\mathbf{W}, \mathbf{H})}$, with the following generator acting on functions $v_i = v_i(t, x)$ with $x \in \mathbb{R}$

$$A^i_t v_i(t, x) = b_i(t, x) \partial_x v_i(t, x) + \frac{1}{2} \sigma^2_i(t, x) \partial^2_{x^2} v_i(t, x) \tag{11}$$

for suitable drift and diffusion coefficients $b_i(t, x)$ and $\sigma_i(t, x)$.

Note that under such a specification of the intensities, dependence between defaults in the model does not only stem from the possibility of common jumps as in Example 2.2 but it can also come from the factor process $\mathbf{X}$ as in Example 2.7 below.

In the above assumption we require that $X^i_t = \varphi_i(\mathbf{X}_t)$ is a Markov process. This assumption is a non-trivial in general, as a process which is a measurable function of a Markov process does not have to be a Markov process itself. We refer to Pitman and Rogers [33] for some discussion of this issue. In our model set-up one, one can show that under appropriate regularity conditions, if for every $(t, \mathbf{x}, x)$ with $\mathbf{x} = (x_Y)_{Y \in \mathcal{Y}}$ and $x = \varphi_i(\mathbf{x})$, one has

$$\begin{aligned} \sum_{\{Y \in \mathcal{Y}\}} b_Y(t, \mathbf{x}) \partial_{x_Y} \varphi_i(\mathbf{x}) &= b_i(t, x) \\ \sum_{\{Y \in \mathcal{Y}\}} \sigma^2_Y(t, \mathbf{x})(\partial_{x_Y} \varphi_i(\mathbf{x}))^2 &= \sigma^2_i(t, x) \end{aligned} \tag{12}$$

then $X_t^i = \varphi_i(\mathbf{X}_t)$ is an $\mathcal{F}$-Markov process with generator A^i in (11). The proof follows from Lemma A.2 (up to the reservation which is made right after the lemma regarding technicalities about the domain of the generators) since for every regular test-function $v_i = v_i(t, x)$, one has with $u(t, \mathbf{x}) := v_i(t, \varphi^i(\mathbf{x}))$

$$v_i(t, X_t^i) - \int_0^t (\partial_s + A_s^i)v_i(s, X_s^i)ds$$
$$= u(t, \mathbf{X}_t) - \int_0^t (\partial_s + A_s)u(s, \mathbf{X}_s)ds.$$

In the two examples given below, the $\mathcal{F}$-Markov property of $X_t^i = \varphi_i(\mathbf{X}_t)$ also rigorously follows, in case of Example 2.6 where φ_i is a coordinate projection operator, from the Markov consistency results of [9], or, in case of Example 2.7, from the semimartingale representation of X^i provided by the SDE (14). The $\mathcal{F}$-Markov property of X^i in Example 2.7 thus follows from the fact that a strong solution to the Markovian SDE (14) driven by the $\mathcal{F}$-Brownian motion W^i, is an $\mathcal{F}$-Markov process, by application of Theorem 32 page 100 of Protter [32]. Example 2.7 is important, as it goes beyond the case of Example 2.6 where the λ_I are deterministic functions of time, and it provides a fully stochastic specification of the λ_Y (including the λ_I).

Example 2.6 (Deterministic Group Intensities) For every group $I \in \mathcal{I}$, the intensity $\lambda_I(t, x_I)$ does not depend on x_I.

Letting $\varphi_i(\mathbf{x}) = x_{\{i\}}$, then (10) and (12) hold with

$$\lambda_i(t, x) := \lambda_{\{i\}}(t, x) + \sum_{\{I \in \mathcal{I}; i \in I\}} \lambda_I(t)$$
$$b_i(t, x) := b_{\{i\}}(t, x)$$
$$\sigma_i(t, x) := \sigma_{\{i\}}(t, x).$$

So, $X^i = X^{\{i\}}$ is $\mathcal{F}$-Markov with drift and diffusion coefficients $b_i(t, x)$ and generator $\sigma_i(t, x)$ thus specified.

Example 2.7 (Extended CIR Intensities) For every $Y \in \mathcal{Y}$, the pre-specified group intensities are given by $\lambda_Y(t, X_t^Y) = X_t^Y$, where the factor X^Y is an extended CIR process

$$(13) \qquad dX_t^Y = a(b_Y(t) - X_t^Y)dt + c\sqrt{X_t^Y}dW_t^Y$$

for non-negative constants a, c and non-negative functions $b_Y(t)$. The SDEs for the factors X^Y have thus the same coefficients except for the $b_Y(t)$.

Letting $\varphi_i(\mathbf{x}) = \sum_{\{Y\in\mathcal{Y};\,i\in Y\}} x_Y = x_{\{i\}} + \sum_{\{I\in\mathcal{I};\,i\in I\}} x_I$, and denoting likewise $b_i(t) = \sum_{\{Y\in\mathcal{Y};\,i\in Y\}} b_Y(t) = b_{\{i\}}(t) + \sum_{\{I\in\mathcal{I};\,i\in I\}} b_I(t)$, then (10) and (12) hold with

$$\lambda_i(t, x) := x$$
$$b_i(t, x) := a(b_i(t) - x)$$
$$\sigma_i(t, x) := c\sqrt{x}.$$

So, $X^i = \sum_{\{Y\in\mathcal{Y};\,i\in Y\}} X^Y$ is an $\mathcal{F}$-Markov process with drift and diffusion coefficients $b_i(t, x)$ and generator $\sigma_i(t, x)$ thus specified.

Note that X^i satisfies the following extended CIR SDE with parameters a, $b_i(t)$ and c as

$$(14) \qquad dX_t^i = a(b_i(t) - X_t^i)dt + c\sqrt{X_t^i}dW_t^i$$

for the $\mathcal{F}$-Brownian motion W^i such that

$$\sqrt{X_t^i}dW_t^i = \sum_{i\in Y} \sqrt{X_t^Y}dW_t^Y, \quad dW_t^i = \sum_{i\in Y} \frac{\sqrt{X_t^Y}}{\sqrt{\sum_{i\in Y} X_t^Y}}dW_t^Y.$$

Remark 2.4. Both the time-deterministic group intensities specification of Example 2.6 and the affine intensities specification of Example 2.7 are used for counterparty credit risk applications in [1, 8, 7] (anticipating the theoretical aspects of the model which are dealt with in the present paper).

The set of obligors alive (resp. in default) at time t is denoted by $J_t = \mathrm{supp}^c(\mathbf{H}_t)$ (resp. $H_t = \mathrm{supp}(\mathbf{H}_t)$). For every $Y \in \mathcal{Y}$ and every set of non-negative constants t_i, we define the quantities $\Lambda_{s,t}^Y$, Λ_t^Y and θ_t^Y as

$$\Lambda_{s,t}^Y = \int_s^t \lambda_Y(s, X_s^Y)ds, \quad \Lambda_t^Y = \Lambda_{0,t}^Y = \int_0^t \lambda_Y(s, X_s^Y)ds \quad \text{and} \quad \theta_t^Y = \max_{i\in Y\cap J_t} t_i$$

where $Y \cap J_t$ in θ_t^Y is the set of survivors in Y at time t (and we use in θ_t^Y our convention that max $\emptyset = -\infty$). Let τ_i denote the default time for obligor i. Since H^i is the default indicator of name i, we have

$$\tau_i = \inf\{t > 0\,;\, H_t^i = 1\}, \quad H_t^i = \mathbb{1}_{\{\tau_i \le t\}}.$$

The following Proposition gathers the Markov properties of the model.

Proposition 2.5. (i) $(\mathbf{X}, \mathbf{H})$ is an $\mathcal{F}$-Markov process with infinitesimal generator given by $\mathcal{A}$.

(ii) *For every obligor i, (X^i, H^i) is an $\mathcal{F}$-Markov process*[2] *admitting the following generator acting on functions $u_i = u_i(t, x_i, k_i)$ with $(x_i, k_i) \in \mathbb{R} \times \{0, 1\}$*

$$\mathcal{A}_t^i u_i(t, x_i, k_i) = b_i(t, x_i)\partial_{x_i} u_i(t, x_i, k_i) + \frac{1}{2}\sigma_i^2(t, x_i)\partial_{x_i^2}^2 u_i(t, x_i, k_i)$$

(15)
$$+\lambda_i(t, x_i)(u_i(t, x_i, 1) - u_i(t, x_i, k_i)).$$

Moreover, the $\mathcal{F}$-intensity process[3] *of H^i is given by $\mathbb{1}_{\{\tau_i > t\}}\lambda_i(t, X_t^i)$. In other words, the process M^i defined by*

(16)
$$M_t^i = \mathbb{1}_{\{\tau_i \le t\}} - \int_0^t \mathbb{1}_{\{\tau_i > s\}}\lambda_i(s, X_s^i)ds,$$

is an $\mathcal{F}$-martingale.[4]

(iii) *For any fixed non-negative constants $t, t_1, \ldots, t_n$, one has*

(17)
$$\mathbb{P}(\tau_1 > t_1, \ldots, \tau_n > t_n \mid \mathcal{F}_t) = \mathbb{P}(\tau_1 > t_1, \ldots, \tau_n > t_n \mid \mathbf{H}_t, \mathbf{X}_t)$$

$$= \mathbb{1}_{\{t_i < \tau_i, \; i \in H_t\}} \mathbb{E}\left\{ \exp\left(-\sum_{Y \in \mathcal{Y}} \Lambda_{t, \theta_t^Y}^Y\right) \middle| \mathbf{X}_t\right\}.$$

The conditional survival probability function of every obligor i is given by, for every $t_i \ge t$,

$$\mathbb{P}(\tau_i > t_i \mid \mathcal{F}_t) = \mathbb{P}(\tau_i > t_i \mid \mathbf{H}_t, \mathbf{X}_t)$$

(18)
$$= \mathbb{1}_{\{\tau_i > t\}} \mathbb{E}\left\{ \exp\left(-\sum_{Y \in \mathcal{Y}, \; i \in Y} \Lambda_{t, t_i}^Y\right) \middle| \mathbf{X}_t\right\}$$

$$= \mathbb{1}_{\{\tau_i > t\}} \mathbb{E}\left\{ \exp\left(-\int_t^{t_i} \lambda_i(s, X_s^i)ds\right) \mid X_t^i\right\}$$

$$= \mathbb{1}_{\{\tau_i > t\}} G_t^i(t_i),$$

with

(19)
$$G_t^i(t_i) = \mathbb{E}\left\{ \exp\left(-\int_t^{t_i} \lambda_i(s, X_s^i)ds\right) \mid X_t^i\right\}.$$

Proof. See Appendix A.2.1.

---

[2] And hence an $\mathcal{F}^{(X^i, H^i)}$-Markov process.

[3] And hence, $\mathcal{F}^{(X^i, H^i)}$-intensity process.

[4] And hence, an $\mathcal{F}^{(X^i, H^i)}$-martingale.

We shall illustrate part (iii) of the above proposition using the following example.

Example 2.3. In case of two obligors and $\mathcal{Y} = \{\{1\}, \{2\}, \{1, 2\}\}$, one can easily check that (17) boils down to

$$
\mathbb{P}(\tau_1 > t_1, \tau_2 > t_2 \mid \mathcal{F}_t) = \mathbb{1}_{\{\tau_1 > t\}} \mathbb{1}_{\{\tau_2 > t\}} \mathbb{E}\left\{ \exp\left(-\sum_{Y \in \mathcal{Y}} \int_t^{t_1 \vee t_2} \lambda_Y(s, X_s^Y)\right) \Big| \mathbf{X}_t \right\}
$$

$$
+ \mathbb{1}_{\{t_2 < \tau_2 \leq t\}} \mathbb{1}_{\{\tau_1 > t\}} \mathbb{E}\left\{ \exp\left(-\int_t^{t_1} \lambda_1(s, X_s^1)\, ds\right) \Big| X_t^1 \right\}
$$

$$
+ \mathbb{1}_{\{t_1 < \tau_1 \leq t\}} \mathbb{1}_{\{\tau_2 > t\}} \mathbb{E}\left\{ \exp\left(-\int_t^{t_2} \lambda_2(s, X_s^2)\, ds\right) \Big| X_t^2 \right\}
$$

$$
+ \mathbb{1}_{\{t_1 < \tau_1 \leq t\}} \mathbb{1}_{\{t_2 < \tau_2 \leq t\}}.
$$

3. Pricing and Hedging Issues

This section treats the pricing, calibration and hedging issues in the Markov copula model of Section 2. First, in Subsection 3.1 we derive the price dynamics for CDS contracts and for CDO tranche in this model. In Subsection 3.2 we use dynamics of Subsection 3.1 to derive min-variance hedging strategies in the Markov copula model.

For notational convenience, we assume zero interest rates. The extension of all theoretical results to time dependent, deterministic interest rates is straightforward but more cumbersome notationally, especially regarding hedging. Time-dependent deterministic interest rates will be used in the numerical part.

3.1 Pricing Equations

In this subsection we derive price dynamics formulas for CDS contracts and CDO tranches in the Markov model; all prices are considered from perspective of the protection buyers. These dynamics will be useful when deriving the min-variance hedging strategies in Subsection 3.2.

In a zero interest-rates environment, the (ex-dividend) price process of an asset is simply given by the risk neutral conditional expectation of future cash flows associated with the asset; the cumulative value process is the sum of the price process and of the cumulative cash-flows process. The cumulative value process is a martingale, as opposed to the price process. When it comes to hedging, the cumulative value process is the main quantity of interest.

For a fixed maturity T, we let S_i denote the T-year CDS spread for obligor i, with recovery rate R_i. Similarly, we let S denote the T-year model CDO tranche spread for the tranche $[a, b]$, with payoff process

(20) $$ L_t^{a,b} = L_{a,b}(\mathbf{H}_t) = (L_t - a)^+ - (L_t - b)^+, $$

where $L_t = \frac{1}{n} \sum_{i=1}^{n} (1 - R_i) H_t^i$ is the credit loss process for the underlying portfolio. The premium legs in these products are payed at $t_1 < t_2 < \ldots < t_p = T$ where $t_j - t_{j-1} =$ h and h is typically a quarter. Below, the notation is the same as in the Itô formula (6).

Proposition 3.1. (i) *The price P^i and the cumulative value $\widehat{P^i}$ at time $t \in [0, T]$ of the single-name CDS on obligor i with contractual spread S_i are given by*

$$P_t^i = v_i(t, \mathbf{X}_t, \mathbf{H}_t)$$
$$(21) \; d\widehat{P_t^i} = \nabla v_i(t, \mathbf{X}_t, \mathbf{H}_t) \sigma(t, \mathbf{X}_t) d\mathbf{W}_t + \sum_{Z \in \mathcal{Z}_t} \mathbf{1}_{i \in Z} \left(1 - R_i - v_i(t, \mathbf{X}_t, \mathbf{H}_t)\right) dM_t^Z$$

for a pre-default pricing function $v_i(t, \mathbf{x}, \mathbf{k})$ such that

$$v_i(t, \mathbf{X}_t, \mathbf{H}_t) = \mathbb{E}[-S_i \text{h} \sum_{t < t_j \leq T} \mathbf{1}_{\{\tau_i > t_j\}} + (1 - R_i) \mathbf{1}_{\{t < \tau_i \leq T\}} | \mathcal{F}_t].$$

(ii) *The price process Π and cumulative value $\widehat{\Pi}$ at time $t \in [0, T]$ of a CDO tranche $[a, b]$ with contractual spread S are given by*

$$\Pi_t = u(t, \mathbf{X}_t, \mathbf{H}_t)$$
$$(22) \quad d\widehat{\Pi}_t = \nabla u(t, \mathbf{X}_t, \mathbf{H}_t) \sigma(t, \mathbf{X}_t) d\mathbf{W}_t$$
$$+ \sum_{Z \in \mathcal{Z}_t} \left(L_{a,b}(\mathbf{H}_{t-}^Z) - L_{a,b}(\mathbf{H}_{t-}) + \delta u^Z(t, \mathbf{X}_t, \mathbf{H}_{t-}) \right) dM_t^Z$$

for a pricing function $u(t, \mathbf{x}, \mathbf{k})$ such that

$$u(t, \mathbf{X}_t, \mathbf{H}_t) = \mathbb{E}\left[-S \text{ h} \sum_{t < t_j \leq T} \left(b - a - L_{t_j}^{a,b} \right) + L_T^{a,b} - L_t^{a,b} \Big| \mathcal{F}_t \right].$$

Proof. See Appendix A.2.2.

Regarding part (i), note that in view of the marginal Markov properties of the model:

- the pricing function $v_i(t, \mathbf{x}, \mathbf{k})$ can essentially be reduced to a "univariate" pre-default pricing function $\tilde{v}_i(t, x_i)$;

- the compensated jump martingale in $\widehat{P^i}$ can be reduced to a "univariate" martingale representation based on the compensated martingale M^i of H^i in (16).

However, as will be clear from Subsection 3.2, the "multivariate" representations of part (i) are more useful in order to handle the hedging issue.

The pricing functions v_i and u can be characterized as the unique solutions to the related Kolmogorov equation (42) in Appendix A.2.2 (or, in the CDS case, a "univariate" Kolmogorov equation can be derived to characterize $\tilde{v}_i$). If the pricing functions are known, the prices at a given time are recovered by plugging the corresponding state of the model into the right-hand-side of the first lines of (21) or (22). But the pricing equation (42) for a CDO tranche leads to a huge system of PDEs which in practice is impossible to handle numerically as soon as n is larger than a few units. As a remedy for this, in Subsection 2.1 of the companion paper [5], we will instead use the translation to a Marshall-Olkin framework which allows us to derive practical recursive pricing schemes for CDO tranche price processes (whereas in practice CDS computations will be based on exponential-affine methodologies).

3.2 Min-Variance Hedging

In this subsection we use the price dynamics from Subsection 3.1 to derive min-variance hedging strategies in the Markov copula model. By min-variance hedging strategies we mean strategies that minimize the variance of the hedging error. Note that in principle one would prefer to minimize the variance relatively to the historical probability measure, however in this paper we minimize the risk-neutral variance for simplicity: see Schweizer [34] for a survey about various quadratic hedging approaches. The hedging strategies are theoretically sound due to our bottom-up Markovian framework and they will be shown in the companion paper [5] to be computationally tractable thanks to the Marshall-Olkin copula interpretation of the model.

Consider a CDO tranche $[a, b]$ with pricing function u specified in Proposition 3.1. Our aim is to find explicit min-variance hedging formulas when hedging this CDO tranche by using the savings account and d single-name CDSs with pricing functions v_i given by Proposition 3.1. First we introduce the CDS cumulative value vector-function

$$\mathbf{v}(t, \mathbf{x}, \mathbf{k}) = (\mathbb{1}_{k_1=0} v_1(t, \mathbf{x}, \mathbf{k}) + \mathbb{1}_{k_1=1}(1 - R_1), \ldots, \mathbb{1}_{k_d=0} v_d t, \mathbf{x}, \mathbf{k}) + \mathbb{1}_{k_d=1}(1 - R_d))^\top .$$

Let $\nabla\mathbf{v}$ denote the Jacobian matrix of $\mathbf{v}$ with respect to $\mathbf{x}$ in the sense of the $d \times v$-matrix such that $\nabla\mathbf{v}(t, \mathbf{x}, \mathbf{k})_i^Y = \partial_{x_Y} v_i(t, \mathbf{x}, \mathbf{k})$, for every $1 \leq i \leq d$ and $Y \in \mathcal{Y}$. Let $\Delta\mathbf{v}^Z$ represent the vector-function of the sensitivities of $\mathbf{v}$ with respect to the event $Z \in \mathcal{N}_n$, so

$$\Delta\mathbf{v}^Z(t, \mathbf{x}, \mathbf{k}) = (\mathbb{1}_{1\in Z, k_1=0}((1-R_1) - v_1(t, \mathbf{x}, \mathbf{k})), \ldots, \mathbb{1}_{d\in Z, k_d=0}((1-R_d)-v_d(t, \mathbf{x}, \mathbf{k})))^\top.$$

By using the vector notation $\widehat{\mathbf{P}} = (\widehat{P^i})_{1 \leq i \leq d}$, one has in view of Proposition 3.1(i)

$$(23) \qquad d\widehat{\mathbf{P}}_t = \nabla\mathbf{v}(t, \mathbf{X}_t, \mathbf{H}_t)\sigma(t, \mathbf{X}_t)d\mathbf{W}_t + \sum_{Z\in\mathcal{Z}_t} \Delta\mathbf{v}^Z(t, \mathbf{X}_t, \mathbf{H}_{t-})dM_t^Z.$$

Let

$$\Delta u^Z(t, \mathbf{x}, \mathbf{k}) = \delta^Z u(t, \mathbf{x}, \mathbf{k}) + L_{a,b}(\mathbf{k}^Z) - L_{a,b}(\mathbf{k})$$

represent the function of sensitivity of the CDO tranche $[a, b]$ cumulative value process with respect to the event $Z \in \mathcal{N}_n$. Let ζ be an d-dimensional row-vector process, representing the number of units held in the first d CDSs which are used in a self-financing[5] hedging strategy for the CDO tranche $[a, b]$. Given (22) and (23), the tracking error (e_t) of the hedged portfolio satisfies $e_0 = 0$ and, for $t \in [0, T]$

$$
\begin{aligned}
(24) \qquad de_t &= d\widehat{\Pi}_t - \zeta_t d\widehat{\mathbf{P}}_t \\
&= \Big(\nabla u(t, \mathbf{X}_t, \mathbf{H}_t) - \zeta_t \nabla \mathbf{v}(t, \mathbf{X}_t, \mathbf{H}_t)\Big) \sigma(t, \mathbf{X}_t) d\mathbf{W}_t \\
&\quad + \sum_{Z \in \mathcal{Z}_t} \Big(\Delta u^Z(t, \mathbf{X}_t, \mathbf{H}_{t-}) - \zeta_t \Delta \mathbf{v}^Z(t, \mathbf{X}_t, \mathbf{H}_{t-})\Big) dM_t^Z.
\end{aligned}
$$

Since the martingale dimension of the model is $\nu + 2^n$, replication is typically out-of-reach[6] in the Markov model. However, in view of (24), we can find min-variance hedging formulas.

Proposition 3.2. *The min-variance hedging strategy ζ is*

$$
(25) \qquad \zeta_t = \frac{d\langle \widehat{\Pi}, \widehat{\mathbf{P}} \rangle_t}{dt} \left(\frac{d\langle \widehat{\mathbf{P}} \rangle_t}{dt} \right)^{-1} = \zeta(t, \mathbf{X}_t, \mathbf{H}_{t-})
$$

where $\zeta = (u, \mathbf{v})(\mathbf{v}, \mathbf{v})^{-1}$, with

$$
\begin{aligned}
(26) \qquad (u, \mathbf{v}) &= (\nabla u)\sigma^2(\nabla \mathbf{v})^T + \sum_{Y \in \mathcal{Y}} \lambda_Y \Delta u^Y (\Delta \mathbf{v}^Y)^T \\
(\mathbf{v}, \mathbf{v}) &= (\nabla \mathbf{v})\sigma^2(\nabla \mathbf{v})^T + \sum_{Y \in \mathcal{Y}} \lambda_Y \Delta \mathbf{v}^Y (\Delta \mathbf{v}^Y)^T.
\end{aligned}
$$

Proof. The first identity in (25) is a classical risk neutral min-variance hedging[7] formula, derived for instance in Section 4.2.3.1 of Crépey [18]. Moreover, one has by computation of the oblique brackets based on the second lines in (21) and (22):

$$
\begin{aligned}
(27) \qquad \frac{d\langle \widehat{\Pi}, \widehat{\mathbf{P}} \rangle_t}{dt} &= \left((\nabla u)\sigma^2(\nabla \mathbf{v})^T + \sum_{Z \in \mathcal{Z}_t} \lambda_Z \Delta u^Z (\Delta \mathbf{v}^Z)^T \right)(t, \mathbf{X}_t, \mathbf{H}_{t-}) = (u, \mathbf{v})(t, \mathbf{X}_t, \mathbf{H}_{t-}) \\
\frac{d\langle \widehat{\mathbf{P}} \rangle_t}{dt} &= \left((\nabla \mathbf{v})\sigma^2(\nabla \mathbf{v})^T + \sum_{Z \in \mathcal{Z}_t} \lambda_Z \Delta \mathbf{v}^Z (\Delta \mathbf{v}^Z)^T \right)(t, \mathbf{X}_t, \mathbf{H}_{t-}) = (\mathbf{v}, \mathbf{v})(t, \mathbf{X}_t, \mathbf{H}_{t-})
\end{aligned}
$$

---

[5] Using also the savings account (constant asset).
[6] See the comments following Proposition 2.2.
[7] See Schweizer [34].

where the second identities in both lines of (27) use simplifications similar to those used in the proof of the Itô formula (6) in Appendix A.1.2. □

In (26), the u-related terms can be computed by using the conditional convolution-recursion procedures developed in the companion paper [5]; the v_i-related terms can be computed very quickly (actually semi-explicitly in either of the specifications of examples 2.6 and 2.7). We will illustrate in [5] the tractability of this approach for computing min-variance hedging deltas.

We refer the reader to Elouerkhaoui [21] for analogous formulas. A nice feature of our set-up however is that due to the specific structure of the intensities, the sums in (26) are over the set $\mathcal{Y}$ of shocks $\mathcal{Y}$ which is of cardinality $v = n + m$, as opposed to the set $\mathcal{N}_n$ of all set-events Z in [21].

We also refer the reader to Frey and Backhaus [23] for other related min-variance hedging formulas.

A. Appendix
A.1 Model Construction

The point process $\mathbf{H}$ with intensity depending on the factor process $\mathbf{X}$ in (3), is constructed by an $\mathbf{X}$-related change of probability measure, starting from an independent continuous-time Markov chain under an auxiliary probability measure $\widehat{\mathbb{P}}$. So, given a factor process $\mathbf{X}$ as in (1) where $\mathbf{W}$ is a $\widehat{\mathbb{P}}$-Brownian motion, let $\mathbf{H}$ denote a continuous-time Markov chain with $\widehat{\mathbb{P}}$-intensity one of transition from $\mathbf{k}$ to $\mathbf{l}$, for every $\mathbf{l} \neq \mathbf{k}$. Let then the $\widehat{\mathbb{P}}$-martingale[8] Γ be defined by $\Gamma_0 = 1$ and, for $t \in [0, T]$,

$$\frac{d\Gamma_t}{\Gamma_{t-}} = \sum_{\mathbf{l} \in \{0,1\}^n} (\lambda(t, \mathbf{X}_t, \mathbf{H}_{t-}, \mathbf{l}) - 1)(dN_t(\mathbf{H}_{t-}, \mathbf{l}) - \mathbb{1}_{\mathbf{l} \neq \mathbf{H}_{t-}} dt)$$
$$= \sum_{\mathbf{l} \neq \mathbf{H}_{t-}} (\lambda(t, \mathbf{X}_t, \mathbf{H}_{t-}, \mathbf{l}) - 1)(dN_t(\mathbf{H}_{t-}, \mathbf{l}) - dt),$$

where the functions $\lambda(t, \mathbf{x}, \mathbf{k}, \mathbf{l})$ are those of (3), and where $N_t(\mathbf{k}, \mathbf{l})$ is the point process with $\widehat{\mathbb{P}}$-intensity $\mathbb{1}_{\{\mathbf{k} = \mathbf{H}_{t-}, \mathbf{l} \neq \mathbf{k}\}}$ counting the transitions of $\mathbf{H}$ from $\mathbf{k}$ to $\mathbf{l}$, for every $\mathbf{k}, \mathbf{l} \in \{0, 1\}^n$. Defining the measure $\mathbb{P}$ by $\frac{d\mathbb{P}}{d\widehat{\mathbb{P}}} = \Gamma_T$, it is then standard to check[9] that the point process $\mathbf{H}$ has intensity (3) under $\mathbb{P}$. To be precise the intensity of $N_t(\mathbf{k}, \mathbf{l})$ is given by (3), with respect to the model filtration $\mathcal{F} = \mathcal{F}^{(\mathbf{W}, \mathbf{H})}$, and the probability measure $\mathbb{P}$. Moreover, process $\mathbf{W}$ remains a Brownian motion under $\mathbb{P}$, the measure-change preserves Markov property of $\mathbf{X}$ with respect to filtration $\mathcal{F}$, and the generator of $\mathbf{X}$ under the new measure is still A_t.

---

[8] Under suitable regularity and growth assumptions on the model coefficients, see Ethier and Kurtz [22] or Crépey [18].

[9] See for instance the proof of Lemma 12.3.5 in Crépey [18].

42

Note that since martingale representation holds under $\widehat{\mathbb{P}}$,[10] martingale representation also holds under the equivalent measure $\mathbb{P}$.

Remark A.1. The prevailing risk neutral probability measure in the paper is $\mathbb{P}$, whereas the auxiliary measure $\widehat{\mathbb{P}}$ is only a mathematical tool used for constructing the model, with no particular financial interpretation.

A.1.1 Proof of Lemma 2.1

By definition of the set-event indicator process H^Z, where $Z \in \mathcal{N}_n$, one has in our model, for $t \in [0, T]$,

$$dH_t^Z = \sum_{\{\mathbf{k},\mathbf{l}\in\{0,1\}^n \,;\, \mathrm{supp}(\mathbf{l})\setminus\mathrm{supp}(\mathbf{k})=Z\}} dN_t(\mathbf{k},\mathbf{l}).$$

So, by (3),

$$\ell_t^Z = \sum_{\{\mathbf{k},\mathbf{l}\in\{0,1\}^n \,;\, \mathrm{supp}(\mathbf{l})\setminus\mathrm{supp}(\mathbf{k})=Z\}} \mathbb{1}_{\{\mathbf{H}_{t-}=\mathbf{k}\}} \sum_{\{Y\in\mathcal{Y};\, \mathbf{k}^Y=1\}} \lambda_Y(t,X_t^Y)$$

$$= \sum_{\{\mathbf{l}\in\{0,1\}^n \,;\, \mathrm{supp}(\mathbf{l})\setminus\mathrm{supp}(\mathbf{H}_{t-})=Z\}} \sum_{\{Y\in\mathcal{Y};\, \mathbf{H}_{t-}^Y=1\}} \lambda_Y(t,X_t^Y)$$

$$= \sum_{\{Y\in\mathcal{Y};\, \mathrm{supp}(\mathbf{H}_{t-}^Y)\setminus\mathrm{supp}(\mathbf{H}_{t-})=Z\}} \lambda_Y(t,X_t^Y)$$

$$= \sum_{\{Y\in\mathcal{Y};\, Y_t=Z\}} \lambda_Y(t,X_t^Y).$$

A.1.2 Proof of Proposition 2.2

Observe that $[M^Y, M^Z] = 0$ for $Y \neq Z$. One thus has the following Itô formula (see for instance Theorem 3.89 page 109 of Jacod [26] or Crépey [18])

$$(28) \quad \begin{aligned} du(t,\mathbf{X}_t,\mathbf{H}_t) &= \big(\partial_t + A_t\big)u(t,\mathbf{X}_t,\mathbf{H}_t)dt + \nabla u(t,\mathbf{X}_t,\mathbf{H}_t)\sigma(t,\mathbf{X}_t)d\mathbf{W}_t \\ &+ \sum_{Z\in\mathcal{N}_n} \delta u^Z(t,\mathbf{X}_t,\mathbf{H}_{t-})dH_t^Z \end{aligned}$$

where we write

$$(29) \quad A_t u(t,\mathbf{x},\mathbf{k}) = \sum_{Y\in\mathcal{Y}}\Big(b_Y(t,x_Y)\partial_{x_Y}u(t,\mathbf{x},\mathbf{k}) + \frac{1}{2}\sigma_Y^2(t,x_Y)\partial_{x_Y^2}^2 u(t,\mathbf{x},\mathbf{k})\Big).$$

Moreover, the structure (4) of the set intensities implies that

$$\sum_{Z\in\mathcal{N}_n} \delta u^Z(t,\mathbf{X}_t,\mathbf{H}_{t-})dH_t^Z = \sum_{Z\in\mathcal{Z}_t} \delta u^Z(t,\mathbf{X}_t,\mathbf{H}_{t-})dH_t^Z,$$

---

[10]In virtue of standard arguments, see for instance Chapter 10 of [27].

which we may further rewrite as

$$\sum_{Z \in \mathcal{Z}_t} \ell_Z(t, \mathbf{X}_t, \mathbf{H}_{t-}) \delta u^Z(t, \mathbf{X}_t, \mathbf{H}_{t-}) dt$$
$$+ \sum_{Z \in \mathcal{Z}_t} \left(\delta u^Z(t, \mathbf{X}_t, \mathbf{H}_{t-}) dH_t^Z - \ell_Z(t, \mathbf{X}_t, \mathbf{H}_t) \delta u^Z(t, \mathbf{X}_t, \mathbf{H}_t) dt \right).$$

Here the second term is $\sum_{Z \in \mathcal{Z}_t} \delta u^Z(t, \mathbf{X}_t, \mathbf{H}_{t-}) dM_t^Z$, whereas one has by (4) in the first term:

$$\sum_{Z \in \mathcal{Z}_t} \ell_Z(t, \mathbf{X}_t, \mathbf{H}_{t-}) \delta u^Z(t, \mathbf{X}_t, \mathbf{H}_{t-})$$
$$= \sum_{Z \in \mathcal{Z}_t} \sum_{Y \in \mathcal{Y}; \, Y_t = Z} \lambda_Y(t, X_t^Y) \delta u^Z(t, \mathbf{X}_t, \mathbf{H}_{t-})$$
$$= \sum_{Y \in \mathcal{Y}} \lambda_Y(t, X_t^Y) \delta u^Y(t, \mathbf{X}_t, \mathbf{H}_{t-})$$

using in the last identity that

$$\delta u^Z(t, \mathbf{x}, \mathbf{k}) = \delta u^Y(t, \mathbf{x}, \mathbf{k}),$$

for every $t, \mathbf{x}, \mathbf{k}, Y$ and Z such that $Y_t = Z$. Thus (28) indeed reduces to (6).

A.2 Markov Properties

Let us first recall the following local martingale characterization of a Markov process with generator $\mathcal{L}$. We work under the standing assumption that uniqueness holds for the solution of the martingale problem defined by $\mathcal{L}$.

Lemma A.2. See, e.g., Ethier and Kurtz [22] *Let X be a right-continuous process with Euclidean state space E, adapted to some filtration $\mathcal{F}$. For X to be an $\mathcal{F}$-Markov process with infinitesimal generator $\mathcal{L}$, it is necessary and sufficient that, for every real-valued function φ in the domain of $\mathcal{L}$,*

$$(30) \qquad \varphi(t, X_t) - \int_0^t (\partial_s + \mathcal{L}_s)\varphi(s, X_s) ds$$

is an $\mathcal{F}$- local martingale.

We shall use this characterization informally in this paper, ignoring the technicalities related to the notion of domain of an operator. Furthermore, throughout the paper we work under the standing assumption that the valuation equation associated to any infinitesimal generator that we use, is well posed in an appropriate functional space. Finally, we assume that uniqueness holds for the solution of the related martingale problem. The reader is referred to Ethier and Kurtz [22] for more details and for specific conditions which can be postulated in these regards.

A.2.1 Proof of Proposition 2.5

(i) In view of the Itô formula (6), $(\mathbf{X}, \mathbf{H})$ solves the martingale problem with generator $\mathcal{A}$ in the filtration $\mathcal{F}$, and is thus an $\mathcal{F}$-Markov process.

(ii) By application of the local martingale characterization of an $\mathcal{F}$-Markov process $(\mathbf{X}, \mathbf{H})$ with generator $\mathcal{A}$ to test-functions of the form $u(t, \mathbf{x}, \mathbf{k}) = v_i(t, x_i, k_i)$, we get the local martingale characterization of an $\mathcal{F}$-Markov process with generator $\mathcal{A}^i$ for (X^i, H^i). Considering $v_i(t, x_i, k_i) = \mathbb{1}_{k_i=1}$ therein yields that M^i in (16) is an $\mathcal{F}$-local martingale.

(iii) We denote $t_Z = \max_{i \in Z} t_i$, for every $Z \in \mathcal{N}_n$. Formula (17) follows directly from Lemma A.3 below since one has, for every $t, t_1, \ldots, t_n \geq 0$,

$$\mathbb{P}(\tau_1 > t_1, \ldots, \tau_n > t_n \mid \mathcal{F}_t) = \sum_{Z \in \mathcal{N}_n} \mathbb{1}_{\{J_t = Z\}} \mathbb{P}(\tau_1 > t_1, \ldots, \tau_n > t_n \mid \mathcal{F}_t)$$

$$= \sum_{Z \in \mathcal{N}_n} \left(\prod_{i \notin Z} \mathbb{1}_{t_i < \tau_i \leq t} \right) \mathbb{E}\left\{ \prod_{i \in Z} \mathbb{1}_{\tau_i > t_i \vee t} \,\middle|\, \mathcal{F}_t \right\}$$

and

$$\mathbb{1}_{\{t_i < \tau_i, \, i \in H_t\}} \mathbb{E}\left\{ \exp\left(-\sum_{Y \in \mathcal{Y}} \Lambda^Y_{t, \theta^Y_t} \right) \,\middle|\, \mathbf{X}_t \right\}$$

$$= \sum_{Z \in \mathcal{N}_n} \mathbb{1}_{\{J_t = Z\}} \mathbb{1}_{\{t_i < \tau_i, \, i \notin Z\}} \mathbb{E}\left\{ \exp\left(-\sum_{Y \in \mathcal{Y}} \Lambda^Y_{t, \theta^Y_t} \right) \,\middle|\, \mathbf{X}_t \right\}$$

$$= \sum_{Z \in \mathcal{N}_n} \left(\prod_{i \notin Z} \mathbb{1}_{t_i < \tau_i \leq t} \right)\left(\prod_{i \in Z} \mathbb{1}_{\tau_i > t} \right) \mathbb{E}\left\{ \exp\left(-\sum_{Y \in \mathcal{Y}} \Lambda^Y_{t, t_{Y \cap Z}} \right) \,\middle|\, \mathbf{X}_t \right\}.$$

Given (17), the other formulas of part (iii) in Proposition 2.5 are straightforward.

Lemma A.3. *For every $t, t_1, \ldots, t_n \geq 0$, and for every $Z \in \mathcal{N}_n$, one has,*

$$(31) \quad \mathbb{E}\left\{ \prod_{i \in Z} \mathbb{1}_{\tau_i > t_i \vee t} \,\middle|\, \mathcal{F}_t \right\} = \left(\prod_{i \in Z} \mathbb{1}_{\tau_i > t} \right) \mathbb{E}\left\{ \exp\left(-\sum_{Y \in \mathcal{Y};\, Y \cap Z \neq \emptyset} \Lambda^Y_{t, t_Y} \right) \,\middle|\, \mathbf{X}_t \right\}.$$

Proof. It is enough to prove that for $t_i \geq t$ one has, for every $Z \in \mathcal{N}_n$,

$$(32) \quad \mathbb{E}\left\{ \prod_{i \in Z} \mathbb{1}_{\tau_i > t_i} \,\middle|\, \mathcal{F}_t \right\} = \left(\prod_{i \in Z} \mathbb{1}_{\tau_i > t} \right) \mathbb{E}\left\{ \exp\left(-\sum_{Y \in \mathcal{Y}} \Lambda^Y_{t, t_{Y \cap Z}} \right) \,\middle|\, \mathbf{X}_t \right\}.$$

Indeed, for general t_i, applying (32) to the $t_i \vee t$ yields

$$\mathbb{E}\left\{ \prod_{i \in Z} \mathbb{1}_{\tau_i > t_i \vee t} \,\middle|\, \mathcal{F}_t \right\} = \left(\prod_{i \in Z} \mathbb{1}_{\tau_i > t} \right) \mathbb{E}\left\{ \exp\left(-\sum_{Y \in \mathcal{Y}} \Lambda^Y_{t, \max_{i \in Y \cap Z} t_i \vee t} \right) \,\middle|\, \mathbf{X}_t \right\}$$

$$= \left(\prod_{i \in Z} \mathbb{1}_{\tau_i > t} \right) \mathbb{E}\left\{ \exp\left(-\sum_{Y \in \mathcal{Y};\, Y \cap Z \neq \emptyset} \Lambda^Y_{t, \max_{i \in Z} t_i} \right) \,\middle|\, \mathbf{X}_t \right\},$$

which is (31). Let us thus show (32) for $t_i \geq t$, by induction on the cardinality d of Z. For $d = 0$, the result is trivial. Assuming the result at rank $d - 1 \geq 0$, let us show the result at rank d. Let us suppose, without loss of generality, that $Z = N_d$ and $t_1 \geq t_2 \geq \cdots \geq t_d \geq t$. One then needs to prove that, using the notation $J^l = 1 - H^l$ for every $l \in N_d$,

$$(33) \qquad \mathbb{E}(\prod_{l=1}^{d} J_{t_l}^l \,|\, \mathcal{F}_t) = (\prod_{l=1}^{d} J_t^l) \mathbb{E}\left\{ \exp\left(- \sum_{Y \in \mathcal{Y}} \Lambda_{t, t_{N_d \cap Y}}^Y \right) \,\middle|\, \mathbf{X}_t \right\}.$$

To establish (33) one first observes that

$$(34) \qquad \mathbb{E}(\prod_{l=1}^{d} J_{t_l}^l \,|\, \mathcal{F}_t) = \mathbb{E}\left\{ J_{t_d}^d \mathbb{E}\left\{ \prod_{l=1}^{d-1} J_{t_l}^l \,\middle|\, \mathcal{F}_{t_d} \right\} \,\middle|\, \mathcal{F}_t \right\},$$

where by the induction hypothesis at rank $d - 1$ the inner conditional expectation can be represented as

$$(35) \qquad (\prod_{l=1}^{d-1} J_{t_d}^l) \mathbb{E}\left\{ \exp\left(- \sum_{Y \in \mathcal{Y}} \Lambda_{t_d, t_{N_{d-1} \cap Y}}^Y \right) \,\middle|\, \mathbf{X}_{t_d} \right\} = (\prod_{l=1}^{d-1} J_{t_d}^l) v(t_d, \mathbf{X}_{t_d})$$

for a suitable function $v = v(t, \mathbf{x})$ over $[0, t_{d-1}] \times \mathbb{R}^{\mathcal{Y}}$, by the Markov property of $\mathbf{X}$. Here the upper bound t_{d-1} for the domain of definition of the function v follows from the fact that $t_d \leq t_{d-1} \leq t_{N_{d-1} \cap Y}$, for every $Y \in \mathcal{Y}$ with $N_{d-1} \cap Y \neq \emptyset$. Inserting (35) into (34) yields by the Markov property of $(\mathbf{X}, \mathbf{H})$ that

$$\mathbb{E}(\prod_{l=1}^{d} J_{t_l}^l \,|\, \mathcal{F}_t) = \mathbb{E}\left\{ (\prod_{l=1}^{d} J_{t_d}^l) v(t_d, \mathbf{X}_{t_d}) \,\middle|\, \mathcal{F}_t \right\} = u(t, \mathbf{X}_t, \mathbf{H}_t),$$

for a function $u = u(t, \mathbf{x}, \mathbf{k})$ over $[0, t_d] \times \mathbb{R}^{\mathcal{Y}} \times \{0, 1\}^n$ characterized by:
$$(36)$$
$$\begin{cases} u(t_d, \mathbf{x}, \mathbf{k}) = (\prod_{l=1}^{d} (1 - k_l)) v(t_d, \mathbf{x}), \quad \mathbf{x} = (x_Y)_{Y \in \mathcal{Y}}, \ \mathbf{k} = (k_1, \ldots, k_n) \in \{0, 1\}^n \\ (\partial_t + \mathcal{A}_t) u(t, \mathbf{x}, \mathbf{k}) = 0, \quad t < t_d, \ \mathbf{x} = (x_Y)_{Y \in \mathcal{Y}}, \ \mathbf{k} \in \{0, 1\}^n. \end{cases}$$

One finally shows that the RHS in (33) admits a representation of the form $(\prod_{l=1}^{d} J_t^l) w(t, \mathbf{X}_t)$, where the function $\tilde{u}(t, \mathbf{x}, \mathbf{k}) = (\prod_{l=1}^{d} (1 - k_l)) w(t, \mathbf{x})$ solves (36). By our standing assumption in this paper equation (36) has a unique solution. Thus $\tilde{u} = u$, which proves (33).

Since $\mathbf{X}$ is Markov with generator A (cf. (i)), the conditional expectation in the RHS of (33) can be represented as $w(t, \mathbf{X}_t)$, for a deterministic function $w = w(t, \mathbf{x})$ over the domain $[0, t_d] \times \mathbb{R}^{\mathcal{Y}}$. In order to get the analytic characterization of w, first note that for every $Y \in \mathcal{Y}$, one has:

$$\Lambda_{t_d, t_{N_{d-1} \cap Y}}^Y = \Lambda_{t_d, t_{N_d \cap Y}}^Y = \Lambda_{t_d, t_{N_d \cap Y}}^Y.$$

This yields the terminal condition $w(t_d, \mathbf{x}) = v(t_d, \mathbf{x})$, $\mathbf{x} = (x_Y)_{Y \in \mathcal{Y}}$. One further has by an application of the Feynman-Kac formula that (see, e.g., Jeanblanc et al. [27])

$$\left(\partial_t + A_t\right)w(t, \mathbf{x}) = \left(\sum_{Y \in \mathcal{Y}\,;\,Y \cap Z \neq \emptyset} \lambda_Y(t, x_Y)\right)w(t, \mathbf{x}), \quad t < t_d, \ \mathbf{x} = (x_Y)_{Y \in \mathcal{Y}}.$$

As a result the function $w = w(t, \mathbf{x})$ is the solution on $[0, t_d] \times \mathbb{R}^{\mathcal{Y}}$ to the following Kolmogorov pricing PDE:

$$(37) \quad \begin{cases} w(t_d, \mathbf{x}) = v(t_d, \mathbf{x}), \quad \mathbf{x} = (x_Y)_{Y \in \mathcal{Y}} \\ \left(\partial_t + A_t\right)w(t, \mathbf{x}) = \left(\sum_{Y \in \mathcal{Y}\,;\,Y \cap Z \neq \emptyset} \lambda_Y(t, x_Y)\right)w(t, \mathbf{x}), \quad t < t_d, \ \mathbf{x} = (x_Y)_{Y \in \mathcal{Y}}. \end{cases}$$

Denoting $\tilde{u}(t, \mathbf{x}, \mathbf{k}) = (\prod_{l \in N_d}(1 - k_l))w(t, \mathbf{x})$, an application of the operator $\mathcal{A}_t$ of (7) yields:

$$\left(\partial_t + \mathcal{A}_t\right)\tilde{u}(t, \mathbf{x}, \mathbf{k}) = (\prod_{l \in N_d}(1 - k_l))\left(\partial_t + A_t\right)w(t, \mathbf{x}) + w(t, \mathbf{x}) \times$$

$$(38) \qquad\qquad \times \sum_{Y \in \mathcal{Y}} \lambda_Y(t, x_Y)\left((\prod_{l \in N_d}(1 - k_l^Y)) - \prod_{l \in N_d}(1 - k_l)\right),$$

where we set, for $Y \in \mathcal{Y}$ and $l \in N_d$,

$$k_l^Y = \begin{cases} 1, & Y \ni l, \\ k_l, & \text{else.} \end{cases}$$

Therefore

$$\sum_{Y \in \mathcal{Y}} \lambda_Y(t, x_Y)\left((\prod_{l \in N_d}(1 - k_l^Y)) - \prod_{l \in N_d}(1 - k_l)\right)$$

$$(39) \qquad\qquad = - \prod_{l \in N_d}(1 - k_l) \sum_{Y \in \mathcal{Y}\,;\,Y \cap N_d \neq \emptyset} \lambda_Y(t, x_Y).$$

Plugging (37) and (39) in the RHS of (38) yields that $(\partial_t + \mathcal{A}_t)\tilde{u}(t, \mathbf{x}, \mathbf{k}) = 0$. Finally $\tilde{u}$ solves (36), which finishes the demonstration. $\qquad\qquad \square$

A.2.2 Proof of Proposition 3.1

Given a function $f = f(t, y)$, let $f(t_j-, x)$ be a notation for the formal limit

$$(40) \qquad\qquad \lim_{(t,y) \to (t_j, x) \text{ with } t < t_j} f(t, y).$$

In view of the Markov property of the model, the following lemma holds in virtue of the Feynman-Kac formula.[11]

Lemma A.4. *Given real-valued functions $\phi(\mathbf{k})$ and $\psi(\mathbf{k})$, one has $\mathbb{E}[\sum_{t < t_j \leq T} \phi(\mathbf{H}_{t_j}) + \psi(\mathbf{H}_T)|\mathcal{F}_t] = w(t, \mathbf{X}_t, \mathbf{H}_t)$, where the function $w(t, \mathbf{x}, \mathbf{k})$ is the*

---

[11] See, e.g., Jeanblanc et al. [27].

solution to the following Kolmogorov pricing PDE system: $w(T, \mathbf{x}, \mathbf{k}) = \psi(\mathbf{k})$, $\mathbf{x} = (x_Y)_{Y \in \mathcal{Y}}$, $\mathbf{k} \in \{0, 1\}^n$, *and for j decreasing from p to 1:*

• *At* $t = t_j$,

$$(41) \qquad w(t_j-, \mathbf{x}, \mathbf{k}) = w(t_j, \mathbf{x}, \mathbf{k}) + \phi(\mathbf{k}), \quad \mathbf{x} = (x_Y)_{Y \in \mathcal{Y}}, \ \mathbf{k} \in \{0, 1\}^n,$$

• *On the time interval* $[t_{j-1}, t_j)$,

$$(42) \qquad \left(\partial_t + \mathcal{A}_t\right) w(t, \mathbf{x}, \mathbf{k}) = 0, \quad \mathbf{x} = (x_Y)_{Y \in \mathcal{Y}}, \ \mathbf{k} \in \{0, 1\}^n.$$

Applying this lemma with

$$\psi_i = (1 - R_i), \quad \phi_i = -S_i \mathrm{h}$$

for part (i) and

$$\psi = L_{a,b}, \quad \phi = -S \mathrm{h}(b - a - L_{a,b})$$

for part (ii) establishes the first lines in identities (21) and (22). Regarding the latter, note that the ex-dividend pricing function $u(t, \mathbf{k}, \mathbf{x})$ in (22), is provided by $w(t, \mathbf{k}, \mathbf{x}) - L_{a,b}(\mathbf{k})$ here.

Moreover, in the filtration $\mathcal{F} = \mathcal{F}^{\mathbf{W}, \mathbf{H}}$, a martingale can only jump at totally unpredictable stopping times. In particular, the cumulative value processes cannot jump at the fixed times t_j. Given the first lines in (21) and (22), the second lines then readily follow using the Itô formula (6).

References

1. Assefa, S., Bielecki, T.R., Crépey, S. and Jeanblanc, M.: CVA computation for counterparty risk assessment in credit portfolios. *Credit Risk Frontiers*, Bielecki, T.R., Brigo, D. and Patras, F., eds., Wiley/Bloomberg-Press, 2011.
2. Basel Committee on Banking Supervision: Revisions to the Basel II market risk framework, December 2010.
3. Bielecki, T.R., Crépey, S., Jeanblanc, M.: Up and down credit risk. *Quantitative Finance* 10 (10), pp. 1137–1151 (2010).
4. Bielecki, T.R., Cousin, A., Crépey, S., Herbertsson, A.: Dynamic Hedging of Portfolio Credit Risk in a Markov Copula Model. Forthcoming in *Journal of Optimization Theory and Applications*.
5. Bielecki, T.R., Cousin, A., Crépey, S., Herbertsson, A.: A Bottom-Up Dynamic Model of Portfolio Credit Risk – Part II: Common-shock interpretation, calibration and hedging issues. In *Recent Advances in Financial Engineering 2012*, World Scientific, 2013.
6. Bielecki, T.R., Cousin, A., Crépey, S., Herbertsson, A.: A bottom-up dynamic model of portfolio credit risk with stochastic intensities and random recoveries, forthcoming in *Communications in Statistics—Theory and Methods*.
7. Bielecki, T. R. and Crépey, S.: Dynamic Hedging of Counterparty Exposure. Forthcoming in *The Musiela Festschrift*, Zariphopoulou, T., Rutkowski, M. and Kabanov, Y., eds, Springer.

8. Bielecki, T.R., Crépey, S., Jeanblanc, M. and Zargari, B.: Valuation and Hedging of CDS Counterparty Exposure in a Markov Copula Model. *International Journal of Theoretical and Applied Finance* 15 (1) 1250004, 2012.

9. Bielecki, T. R. and Jakubowski, J. and Niewęglowski, M.: Dynamic Modeling of Dependence in Finance via Copulae Between Stochastic Processes, *Copula Theory and Its Applications,* Lecture Notes in Statistics, Vol.198, Part 1, 33–76, 2010.

10. Bielecki, T.R., Vidozzi, A. and Vidozzi, L.: Collateralized CVA Valuation with Rating Triggers and Credit Migrations, *J. of Credit Risk*, submitted, 2012.

11. Bielecki, T.R., Cialenco, I. and Iyigunler, I.: A Markov Copulae Approach to Pricing and Hedging of Credit Index Derivatives and Ratings Triggered Step–Up Bonds, *J. of Credit Risk*, 2008.

12. Brigo, D., Pallavicini, A., Torresetti, R.: Calibration of CDO Tranches with the dynamical Generalized-Poisson Loss model. *Risk Magazine*, May 2007.

13. Brigo, D., Pallavicini, A., Torresetti, R. Credit models and the crisis: default cluster dynamics and the generalized Poisson loss model, *Journal of Credit Risk,* 6 (4), 39–81, 2010.

14. Brigo, D., Pallavicini, A., Torresetti, R. Cluster-based extension of the generalized poisson loss dynamics and consistency with single names. *International Journal of Theoretical and Applied Finance,* Vol 10, n. 4, 607-632, 2007.

15. Cont, R. and Kan, Y.H.: Dynamic Hedging of Portfolio Credit Derivatives, *SIAM Journal on Financial Mathematics*, 2, p. 112-140, 2011.

16. Cont, R., Minca, A.: Recovering Portfolio Default Intensities Implied by CDO Quotes, *Mathematical Finance* online first 22 June 2011.

17. Cousin, A., Crépey, S. and Kan, Y.H.: Delta-hedging Correlation Risk? *Review of Derivatives Research*, June 2011.

18. Crépey, S.: *Financial Modeling: A Backward Stochastic Differential Equations Perspective*, Springer, 2013.

19. Crépey, S. and Rahal, A.: Simulation/Regression Pricing Schemes for CVA Computations on CDO Tranches, forthcoming in *Communications in Statistics—Theory and Methods.*

20. Duffie, D. and Gârleanu, N.: Risk and the valuation of collateralized debt obligations, *Financial Analysts Journal*, 57, 41-62, 2001.

21. Elouerkhaoui, Y.: Pricing and Hedging in a Dynamic Credit Model. *International Journal of Theoretical and Applied Finance*, Vol. 10, Issue 4, 703–731, 2007.

22. Ethier, H.J. and Kurtz, T.G.: *Markov Processes. Characterization and Convergence.* Wiley, 1986.

23. Frey, R., Backhaus, J. Dynamic hedging of synthetic CDO tranches with spread- and contagion risk, *Journal of Economic Dynamics and Control,* 34 (4), 710–724, 2010.

24. Herbertsson, A. (2011) Modelling default contagion using Multivariate Phase-Type distributions, *Review of Derivatives Research* 14 (1), 1–36.

25. Herbertsson, A. and Rootzén, H. (2008) Pricing kth-to-default swaps under default contagion:the matrix-analytic approach, *The Journal of Computational Finance* 12 (1), 49–78.

26. JACOD, J.: *Calcul Stochastique et Problèmes de Martingales.* Springer, 2nd edition, 2003.

27. JEANBLANC, M., YOR, M. AND CHESNEY, M.: *Mathematical methods for Financial Mar-*

kets. Springer, 2009.

28. LAURENT, J.-P., COUSIN, A. AND FERMANIAN, J-D.: Hedging default risks of CDOs in Markovian contagion models. Quantitative Finance, 1469-7696, April 2010.

29. Lindskog, F. and McNeil, A. J.: Common Poisson shock models: applications to insurance and credit risk modelling. *ASTIN Bulletin*, 33(2), 209-238, 2003

30. Marshall, A. & Olkin, I.: A multivariate exponential distribution, *J. Amer. Statist. Assoc.*, 2, 84-98, 1967.

31. Patton, A.: Modelling Time-varying exchange rate dependence using the conditional copula, *Working Paper* 2001-09, University of California, San Diego, 2001.

32. Protter, P.E.: *Stochastic Integration and Differential Equations, Second Edition, Version 2.1.* Springer, 2005.

33. Rogers, L.C.G. and Pitman J.W.: Markov functions, *Ann. of Prob.*, 9, 578-582, 1981.

34. Schweizer, M.: A Guided Tour through Quadratic Hedging Approaches. In *Option Pricing, Interest Rates and Risk Management*, E. Jouini, J. Cvitanic and M. Musiela, eds., Cambridge University Press, 538–574, 2001.

A Bottom-Up Dynamic Model of Portfolio Credit Risk. Part II: Common-Shock Interpretation, Calibration and Hedging Issues

Tomasz R. Bielecki[1]*, Areski Cousin[2]†, Stéphane Crépey[3]‡
Alexander Herbertsson[4]§

[1] Department of Applied Mathematics, Illinois Institute of Technology, Chicago, IL 60616, USA [2] Université de Lyon, Université Lyon 1, LSAF, France
[3] Laboratoire Analyse et Probabilités, Université d'Évry Val d'Essonne, 91037 Évry Cedex, France
[4] Centre for finance/Department of Economics, University of Gothenburg, SE 405 30 Göteborg, Sweden

In this paper, we prove that the conditional dependence structure of default times in the Markov model of [4] belongs to the class of Marshall-Olkin copulas. This allows us to derive a factor representation in terms of "common-shocks", the latter beeing able to trigger simultaneous defaults in some pre-specified groups of obligors. This representation depends on the current default state of the credit portfolio so that fast convolution pricing schemes can be exploited for pricing and hedging credit portfolio derivatives. As emphasized in [4], the innovative breakthrough of this dynamic bottom-up model is a suitable decoupling property between the dependence structure and the default marginals as in [9] (like in static copula models but here in a full-flesh dynamic "Markov copula" model). Given the fast deterministic pricing schemes of the present paper, the model can then be jointly calibrated to single-name and portfolio data in two steps, as opposed to a global joint optimization procedures involving all the model parameters at the same time which would be untractable numerically. We illustrate this numerically by results of calibration against market data from CDO tranches as well as individual CDS spreads. We also discuss hedging sensitivities computed in the models thus calibrated.

Key words: Portfolio credit risk, Basket credit derivatives, Markov copula model, Common shocks, Pricing, Calibration, Min-variance hedging.

52

1. Introduction

In [4] we introduced a Markov copula model of default times providing a decoupling between the single-name marginals and the dependence structure of the default times. In this sense, this model solves the portfolio credit risk top-down bottom-up puzzle [7]. For earlier partial progress in this direction, see [10, 11, 14] and the discussion in [4]. This paper is the applicative companion to [4] (see also [3] for a short announcing version of both papers and [5] for further features of the model).

The paper is organized as follows. In Section 2 we provide an "executive summary" of the dynamic model of [4] in the form of an equivalent common-shock representation. In terms of common-shock representation, we mean that each individual default process can be represented by Cox processes likely to trigger defaults simultaneously in some pre-specifed group of obligors. In Section 3 we use the resulting representation to derive fast deterministic pricing algorithms. As emphasized in [4], the innovative breakthrough of the model is a suitable decoupling property between the dependence structure and the individual names [9], like in static copula models but here in a full-flesh dynamic "Markov copula" model. Given the fast deterministic pricing schemes of this paper, the model can then be jointly calibrated to single-name and portfolio data in two steps, as opposed to a global joint optimization procedures involving all the model parameters at the same time which would be untractable numerically. In particular, this model allows one to address in a dynamic and theoretically consistent way the issues of hedging basket credit derivatives by individual names, whilst preserving the static common factor tractability. To illustrate these features, Section 4 presents numerical results of calibration against market data from CDO tranches as well as individual CDS spreads and Section 5 discusses hedging sensitivities computed in the models thus calibrated. Note that this model can be applied well the space of consistent valuation and hedging of CDSs and CDOs. In particular it is used in [2, 8, 13] (see also [12]) for valuation and hedging of counterparty risk on credit derivatives.

In the rest of the paper we consider a risk neutral pricing model $(\Omega, \mathcal{F}, \mathbb{P})$, for a filtration $\mathcal{F} = (\mathcal{F}_t)_{t \in [0,T]}$ which will be specified below and where $T \geq 0$ is a

---

*The research of T.R. Bielecki was supported by NSF Grant DMS–0604789 and NSF Grant DMS–0908099.

†The research of A. Cousin benefited from the support of the DGE and the ANR project Ast&Risk.

‡The research of Stéphane Crépey benefited from the support of the "Chair Markets in Transition" under the aegis of Louis Bachelier laboratory, a joint initiative of École polytechnique, Université d'Évry Val d'Essonne and Fédération Bancaire Française.

§The research of A. Herbertsson was supported by the Jan Wallander and Tom Hedelius Foundation and by Vinnova.

fixed time horizon. We denote $N_n = \{1, \ldots, n\}$ and let $\mathcal{N}_n$ denote the set of all subsets of N_n where n represents the number of obligors in the underlying credit portfolio.

2. Model of Default Times

In the Markov copula common shocks model, defaults are the consequence of some "shocks" associated with groups of obligors. We define the following pre-specified set of groups

$$\mathcal{Y} = \{\{1\}, \ldots, \{n\}, I_1, \ldots, I_m\},$$

where $I_1, \ldots, I_m$ are subsets of $\{1, \ldots, n\}$, and each group I_j contains at least two obligors or more. The shocks are divided in two categories: the "idiosyncratic" shocks associated with singletons $\{1\}, \ldots, \{n\}$ can only trigger the default of name $1, \ldots, n$ individually, while the "systemic" shocks associated with multi-name groups $I_1, \ldots, I_m$ may simultaneously trigger the default of all names in these groups. Note that several groups I_j may contain a given name i, so that only the shock occurring first effectively triggers the default of that name. As a result, when a shock associated with a specific group occurs at time t, it only triggers the default of names that are still alive in that group at time t. In the following, the elements Y of $\mathcal{Y}$ will be used to designate shocks and we let $\mathcal{I} = (I_l)_{1 \le l \le m}$ denote the pre-specified set of multi-name groups of obligors. Shock intensities $\lambda_Y(t, \mathbf{X}_t)$ will be specified later in terms of a Markovian factor process $\mathbf{X}_t$. Letting $\Lambda_t^Y = \int_0^t \lambda_Y(s, \mathbf{X}_s) ds$, we define

$$(1) \qquad \qquad \widehat{\tau}_Y = \inf\{t > 0; \Lambda_t^Y > E_Y\},$$

where the random variables E_Y are i.i.d. and exponentially distributed with parameter 1. For every obligor i we let

$$(2) \qquad \qquad \tau_i = \min_{\{Y \in \mathcal{Y}; i \in Y\}} \widehat{\tau}_Y,$$

which defines the default time of obligor i in the common shocks model. The model filtration is given as $\mathbb{F} = \mathbb{X} \vee \mathbb{H}$, the filtration generated by the factor process $\mathbf{X}$ and the point process $\mathbf{H} = (H^i)_{1 \le i \le n}$ with $H_t^i = \mathbb{1}_{\tau_i \le t}$.

This model can be viewed as a doubly stochastic (via the stochastic intensities Λ^Y) and dynamized (via the introduction of the filtration $\mathbb{F}$) generalization of the Marshall-Olkin model [18]. The purpose of the factor process $\mathbf{X}$ is to more realistically model diffusive randomness of credit spreads. Note that in [4], we present the model the reverse way round, i.e. we first construct a suitable Markov process $(\mathbf{X}_t, \mathbf{H}_t)$ and then define the τ_i as the jump times of the H^i.

The set of obligors alive (resp. in default) at time t is denoted by $J_t = \text{supp}^c(\mathbf{H}_t)$ (resp. $H_t = \text{supp}(\mathbf{H}_t)$). For every $Y \in \mathcal{Y}$ and every set of nonnegative constants $t, t_1, \ldots, t_n$, we define

$$\theta_t^Y = \max_{i \in Y \cap J_t} t_i$$

(with the convention that max $\emptyset = -\infty$). Note that $Y \cap J_t$ in θ_t^Y represents the set of survivors in Y at time t. We also write

(3) $$\Lambda_{s,t}^Y = \int_s^t \lambda_Y(s, X_s^Y)ds, \quad \lambda_t^i = \sum_{\{Y \in \mathcal{Y}; i \in Y\}} \lambda_Y(t, \mathbf{X}_t).$$

The following result (see [3, 4]) is key in the model.

Proposition 2.1. *For any fixed nonnegative constants* $t, t_1, \ldots, t_n$, *we have:*

(4) $$\mathbb{P}(\tau_1 > t_1, \ldots, \tau_n > t_n \mid \mathcal{F}_t) = \mathbb{1}_{\{t_i < \tau_i, \ i \in H_t\}} \mathbb{E}\left\{\exp\left(-\sum_{Y \in \mathcal{Y}} \Lambda_{t,t \vee \theta_t^Y}^Y\right) \Big| \mathbf{X}_t\right\}.$$

In particular, for every obligor i *and* $t_i \geq t$,

(5) $$\mathbb{P}(\tau_i > t_i \mid \mathcal{F}_t) = \mathbb{1}_{\{\tau_i > t\}} \mathbb{E}\left\{\exp\left(-\int_t^{t_i} \lambda_s^i ds\right) \mid \mathbf{X}_t\right\}.$$

In Section 4, we will see that thanks to formula (4) (resp. (5) and under an additional affine structure postulated below on each individual pre-default intensity process λ_t^i), efficient convolution recursion procedures (resp. affine methodologies) are available for pricing multi-name credit derivatives like CDO tranches (resp. single-name credit derivatives like CDSs), conditionally on any given state of the dynamic model $(\mathbf{X}_t, \mathbf{H}_t)$. The model can then be calibrated in two steps: individual λ^i-parameters are first calibrated to individual CDSs and the model dependence λ_I-parameters are then calibrated to CDO tranches (as opposed to a global joint optimization procedures involving all the model parameters at the same time, which would be untractable numerically). But first, as announced above, in order to ensure the Markov consistency and Markov copula feature of the setup (see [9]), we assume further that every individual process λ^i is an affine process (in particular, a Markov process), as in either specification below. Consequently the conditioning with respect to $\mathbf{X}_t$ can be replaced by a conditioning with respect to λ_t^i in (5), hence exponential-affine methodologies for computing (5) follow.

Example 2.2. (i) (Deterministic group intensities). The idiosyncratic intensities $\lambda_{\{i\}}(t, \mathbf{X}_t)$ are affine, and the systemic intensities $\lambda_Y(t, \mathbf{X}_t)$ are deterministic functions of time, i.e. the functions $\lambda_Y(t, \mathbf{x})$ do not depend on $\mathbf{x}$, for $Y \in \mathcal{Y}$ that are not singletons.

(ii) (Extended CIR intensities). $\mathbf{X}_t = (X_t^Y)_{Y\in\mathcal{Y}}$ and for every $Y \in \mathcal{Y}$, $\lambda_Y(t, \mathbf{X}_t) = X_t^Y$, where X_t^Y is an extended CIR process

$$dX_t^Y = a(b_Y(t) - X_t^Y)dt + c\sqrt{X_t^Y}dW_t^Y, \tag{6}$$

for nonnegative constants a, c (independent of Y) and a nonnegative function $b_Y(t)$, and where the W^Y are independent standard Brownian motions.

In the second specification, affinity of λ^i (which is trivial in the first specification) arises from the fact that the SDE for the factors X^Y have the same coefficients except for the $b_Y(t)$. Thus, $X^i := \sum_{\{Y\in\mathcal{Y};\, i\in Y\}} X^Y$ satisfies the following extended CIR SDE:

$$dX_t^i = a(b_i(t) - X_t^i)dt + c\sqrt{X_t^i}dW_t^i, \tag{7}$$

for the function $b_i(t) = \sum_{\{Y\in\mathcal{Y};\, i\in Y\}} b_Y(t)$ and the Brownian motion

$$dW_t^i = \sum_{i\in Y} \frac{\sqrt{X_t^Y}}{\sqrt{\sum_{i\in Y} X_t^Y}}dW_t^Y.$$

3. Fast Deterministic Pricing Schemes
3.1 Common Shocks Model Interpretation

In view of formula (4), conditionally on any given state $(\mathbf{x}, \mathbf{k})$ of $(\mathbf{X}, \mathbf{H})$ at time t, it is possible to define a "conditional common shock model" of default times of the surviving names at time t, such that the law of the default times in the conditional common shock model is the same as the corresponding conditional distribution in the original model. This representation will be used in the next section for deriving fast exact convolution recursion procedures for pricing portfolio loss derivatives.

We thus introduce a family of common shocks copula models, parameterized by the current time t. For every $Y \in \mathcal{Y}$, we define

$$\tau_Y(t) = \inf\{\theta > t; \Lambda_\theta^Y > \Lambda_t^Y + E_Y\},$$

where the random variables E_Y are i.i.d. and exponentially distributed with parameter 1. For every obligor i we let

$$\tau_i(t) = \min_{\{Y\in\mathcal{Y};\, i\in Y\}} \tau_Y(t), \tag{8}$$

which defines the default time of obligor i in the common shocks copula model starting at time t. We also introduce the indicator processes $H_\theta^Y(t) = \mathbb{1}_{\{\tau_Y(t)\leq\theta\}}$ and

$H_\theta^i(t) = \mathbb{1}_{\{\tau_i(t) \le \theta\}}$, for every shock Y, obligor i and time horizon $\theta \ge t$. Let $Z \in \mathcal{N}_n$ denote a set of obligors, meant in the probabilistic interpretation to represent the set J_t of survived obligors in the original model at time t. We now prove that on $\{J_t = Z\}$, the conditional law of $(\tau_i)_{i \in J}$ given $\mathcal{F}_t$ in the original model, is equal to the conditional law of $(\tau_i(t))_{i \in Z}$ given $\mathbf{X}_t$ in the common shocks framework starting at time t. Let also $N_\theta = \sum_{1 \le i \le n} H_\theta^i$ denote the cumulative number of defaulted obligors in the original model up to time θ. Let $N_\theta(t, Z) = n - |Z| + \sum_{i \in Z} H_\theta^i(t)$, denote the cumulative number of defaulted obligors in the time-t common shocks framework up to time θ where $|Z|$ is the cardinality of the set Z.

Proposition 3.1. *Let $Z \in \mathcal{N}_n$ denote an arbitrary subset of obligors and let $t \ge 0$. Then,*
(i) for every $t_1, \ldots, t_n \ge t$, one has

$$(9) \qquad \mathbb{1}_{\{J_t = Z\}} \mathbb{P}\left(\tau_i > t_i, \ i \in J_t \,\middle|\, \mathcal{F}_t\right) = \mathbb{1}_{\{J_t = Z\}} \mathbb{P}\left(\tau_i(t) > t_i, \ i \in Z \,\middle|\, \mathbf{X}_t\right).$$

(ii) for every $\theta \ge t$, one has that for every $k = n - |Z|, \ldots, n$,

$$\mathbb{1}_{\{J_t = Z\}} \mathbb{P}\left(N_\theta = k \,\middle|\, \mathcal{F}_t\right) = \mathbb{1}_{\{J_t = Z\}} \mathbb{P}\left(N_\theta(t, Z) = k \,\middle|\, \mathbf{X}_t\right).$$

Proof. Part (ii) readily follows from part (i), that we now show. Let, for every obligor i, $\tilde{t}_i = \mathbb{1}_{i \in J_t} t_i$. Note that one has, for $Y \in \mathcal{Y}$

$$\max_{i \in Y \cap J_t} \tilde{t}_i = \max_{i \in Y \cap J_t} t_i = \theta_t^Y.$$

Thus, by application of formula (4) to the sequence of times $(\tilde{t}_i)_{1 \le i \le n}$, it comes,

$$\mathbb{1}_{\{J_t = Z\}} \mathbb{P}\left(\tau_i > t_i, \ i \in J_t \,\middle|\, \mathcal{F}_t\right)$$
$$= \mathbb{1}_{\{J_t = Z\}} \mathbb{P}\left((\tau_i > t_i, \ i \in Z), \ (\tau_i > 0, \ i \in Z^c) \,\middle|\, \mathcal{F}_t\right)$$
$$= \mathbb{1}_{\{J_t = Z\}} \mathbb{E}\left\{\exp\left(-\sum_{Y \in \mathcal{Y}} \Lambda_{t, \theta_t^Y}^Y\right) \,\middle|\, \mathbf{X}_t\right\}$$

which on $\{J_t = Z\}$ coincides with the expression

$$\mathbb{E}\left\{\exp\left(-\sum_{Y \in \mathcal{Y}} \Lambda_{t, \max_{i \in Y \cap Z} t_i}^Y\right) \,\middle|\, \mathbf{X}_t\right\}$$

deduced from (4) for $\mathbb{P}(\tau_i(t) > t_i, \ i \in Z \,\middle|\, \mathbf{X}_t)$. $\qquad\square$

3.2 Recursive Convolutive Pricing Schemes

In this subsection we use the conditional common shock model representation to derive fast convolution recursion algorithms for computing the conditional portfolio loss distribution. In the case where the recovery rate is the same for all names, i.e., $R_i = R$, $i = 1, \ldots, n$, the price process for a CDO tranche $[a, b]$ is determined by the probabilities $\mathbb{P}[N_\theta = k \mid \mathcal{F}_t]$ for $k = |\mathbf{H}_t|, \ldots, n$ and $\theta \geq t \geq 0$. But we know from Proposition 3.1(ii) that

$$\mathbb{P}[N_\theta = k \mid \mathcal{F}_t] = \mathbb{P}[N_\theta(t, Z) = k \mid \mathbf{X}_t]$$

on the event $\{J_t = Z\}$, so we will focus on computation of the latter probabilities, which are derived in formula (11) below.

We henceforth assume a nested structure of the sets I_j given by

$$(10) \qquad\qquad I_1 \subset \ldots \subset I_m.$$

This structure implies that if all obligors in group I_k have defaulted, then all obligors in group $I_1, \ldots, I_{k-1}$ have also defaulted. As we shall detail in Remark 3.3, the nested structure (10) yields a particularly tractable expression for the portfolio loss distribution. This nested structure also makes sense financially with regards to the hierarchical structure of risks which is reflected in standard CDO tranches.

Remark 3.2. A dynamic group structure would be preferable from a financial point of view. In the same vein one could deplore the absence of direct contagion effects in this model (it only has joint defaults). However it should be stressed that we are building a pricing model, not an econometric model; the applications we have in mind are hedging CDO tranches by individual names (see Section 5), as well as valuation and hedging of counterparty risk on credit portfolios (see [6]). In these regards, efficient pricing (at any future point in time, not only at time 0 [6]) and Greeking procedures, as well as efficient joint calibration to CDS and CDO data (see Section 4), are the main issues, and these are already quite difficult to achieve simultaneously in a single model.

Denoting conventionally $I_0 = \emptyset$ and $H_\theta^{I_0}(t) = 1$, then in view of (10), the events

$$\Omega_\theta^j(t) := \{H_\theta^{I_j}(t) = 1, H_\theta^{I_{j+1}}(t) = 0, \ldots, H_\theta^{I_m}(t) = 0\}, \;\; 0 \leq j \leq m$$

form a partition of Ω. Hence, we have

$$(11) \qquad \mathbb{P}(N_\theta(t, Z) = k \mid \mathbf{X}_t) = \sum_{0 \leq j \leq m} \mathbb{P}(N_\theta(t, Z) = k \mid \Omega_\theta^j(t), \mathbf{X}_t)\mathbb{P}(\Omega_\theta^j(t) \mid \mathbf{X}_t)$$

where, by construction of the $H_\theta^I(t)$ and independence of the $\lambda_l(t, X_t^l)$ we have

$$(12) \qquad \mathbb{P}(\Omega_\theta^j(t) \mid \mathbf{X}_t) = \left(1 - \mathbb{E}(e^{-\Lambda_{t,\theta}^{I_j}} \mid X_t^{I_j})\right) \prod_{j+1 \leq l \leq m} \mathbb{E}(e^{-\Lambda_{t,\theta}^{I_l}} \mid X_t^{I_l})$$

which in our model can be computed very quickly (actually, semi-explicitly in either of the specifications of Example (2.2)). We now turn to the computation of the term

$$\mathbb{P}(N_\theta(t, Z) = k \mid \Omega_\theta^j(t), \mathbf{X}_t) \tag{13}$$

appearing in (11). Recall first that $N_\theta(t, Z) = n - |Z| + \sum_{i \in Z} H_\theta^i(t)$ with $|Z|$ denoting the cardinality of Z. We know that for every group $j = 1, \ldots, m$, given $\Omega_\theta^j(t)$, the marginal default indicators $H_\theta^i(t)$ for $i \in Z$ are such that:

$$H_\theta^i(t) = \begin{cases} 1, & i \in I_j, \\ H_\theta^{\{i\}}(t), & \text{else.} \end{cases} \tag{14}$$

Hence, the $H_\theta^i(t)$ are conditionally independent given $\Omega_\theta^j(t)$. Finally, conditionally on $(\Omega_\theta^j(t), \mathbf{X}_t)$ the random vector $\mathbf{H}_\theta(t) = (H_\theta^i(t))_{i \in N_n}$ is a vector of independent Bernoulli random variables with parameter $p = (p_\theta^{i,j}(t))_{i \in N_n}$, where

$$p_\theta^{i,j}(t) = \begin{cases} 1, & i \in I_j, \\ 1 - \mathbb{E}\left\{\exp\left(-\Lambda_{t,\theta}^{\{i\}}\right) \mid X_t^{\{i\}}\right\}, & \text{else} \end{cases} \tag{15}$$

The conditional probability (13) can therefore be computed by a standard convolution recursive procedure (see, for instance, Andersen and Sidenius [1]).

Remark 3.3. The linear number of terms in the sum of (11) is due to the nested structure of the groups I_j in (11). Note that a convolution recursion procedure is possible for an arbitrary structuring of the groups I_j. However, a general structuring of the m groups I_j would imply 2^m terms instead of m in the sum of (11), which in practice would only work for very few groups m. The nested structure (11) of the I_j, or equivalently, the tranched structure of the $I_j \setminus I_{j-1}$, is also quite natural from the point of view of application to CDO tranches.

4. Model Calibration

In this section we briefly discuss the calibration of the model and some few numerical results connected to the loss-distributions. Subsection 4.1 outlines the calibration methodology with piecewise constant default intensities and constant recoveries (for a calibration with stochastic intensities and/or random recoveries see in [5]). Then Subsection 4.2 presents the numerical calibration of the Markov copula model against market data. We also study the implied loss-distributions in our fitted model for the case with constant recoveries.

4.1 Calibration Methodology

In this subsection we discuss one of the calibration methodologies that will be used when fitting the Markov copula model against CDO tranches on the iTraxx

Europe and CDX.NA.IG series in Subsection 4.2. This first calibration methodology will use piecewise constant default intensities and constant recoveries in the convolution pricing algorithm of Subsection 3.

The first step is to calibrate the single-name CDS for every obligor. Given the T-year market CDS spread S_i^* for obligor we want to find the individual default parameters for obligor i so that $P_0^i(S_i^*) = 0$, so

$$(16) \qquad S_i^* = \frac{(1 - R_i)\mathbb{P}(\tau_i < T)}{h \sum_{0 < t_j \leq T} \mathbb{P}(\tau_i > t_j)}$$

where we used the facts that interest rate is zero and that the recovery R_i is constant. Hence, the first step is to extract the implied hazard function $\Gamma_i^*(t) = -\ln \mathbb{P}(\tau_i > T)$ from the CDS curve of every obligor i by using a standard bootstrapping procedure based on (16).

Given the marginal hazard functions, the law of the total number of defaults at a fixed horizon is a function of the joint default intensity functions $\lambda_I(t)$, as described by the recursive algorithm of Subsection 3. The second step is therefore to calibrate the common-shock intensities $\lambda_I(t)$ so that the model CDO tranche spreads coincide with the corresponding market spreads. This is done by using the recursive algorithm of Subsection 3, for $\lambda_I(t)$s parameterized as non-negative and piecewise constant functions of time. Moreover, in view of the definition of λ_t^i in (3), for every obligor i and at each time t we impose the constraint

$$(17) \qquad \sum_{I \in \mathcal{I}; i \in I} \lambda_I(t) \leq \lambda_i^*(t)$$

where $\lambda_i^* := \frac{d\Gamma_i^*}{dt}$ denotes the hazard rate (or hazard intensity) of name i. For constant joint default intensities $\lambda_I(t) = \lambda_I$ the constraints (17) reduce to

$$\sum_{I \ni i} \lambda_I \leq \underline{\lambda}_i := \inf_{t \in [0,T]} \lambda_i^*(t) \quad \text{for every obligor } i.$$

Given the nested structure of the groups I_j-s specified in (10), this is equivalent to

$$(18) \qquad \sum_{j=l}^{m} \lambda_{I_j} \leq \underline{\lambda}_{I_l} := \min_{i \in I_l \setminus I_{l-1}} \underline{\lambda}_i \text{ for every group } l.$$

Furthermore, for piecewise constant common shock intensities on a time grid (T_k), the condition (18) extends to the following constraint

$$(19) \quad \sum_{j=l}^{m} \lambda_{I_j}^k \leq \underline{\lambda}_{I_l}^k := \min_{i \in I_l \setminus I_{l-1}} \underline{\lambda}_i^k \quad \text{for every } l, k \quad \text{where } \underline{\lambda}_i^k := \inf_{t \in [T_{k-1}, T_k]} \lambda_i^*(t).$$

We remark that insisting on calibrating all CDS names in the portfolio, including the safest ones, implies via (18) or (19) a very constrained region for the common shock parameters. This region can be expanded by relaxing the system of constraints for the joint default intensities, by excluding the safest CDSs from the calibration.

In this paper we will use a time grid consisting of two maturities T_1 and T_2. Hence, the single-name CDSs constituting the entities in the credit portfolio are bootstrapped from their market spreads for $T = T_1$ and $T = T_2$. This is done by using piecewise constant individual default intensity λ_i-s on the time intervals $[0, T_1]$ and $[T_1, T_2]$.

Before we leave this subsection, we give some few more details on the calibration of the common shock intensities for the m groups in the second calibration step. From now on we assume that the joint default intensities $\{\lambda_{I_j}(t)\}_{j=1}^m$ are piecewise constant functions of time, so that $\lambda_{I_j}(t) = \lambda_{I_j}^{(1)}$ for $t \in [0, T_1]$ and $\lambda_{I_j}(t) = \lambda_{I_j}^{(2)}$ for $t \in [T_1, T_2]$ and for every group j. Next, the joint default intensities $\lambda = (\lambda_{I_j}^{(k)})_{j,k} = \{\lambda_{I_j}^{(k)} : j = 1, \ldots, m \text{ and } k = 1, 2\}$ are then calibrated so that the five-year model spread $S_{a_l, b_l}(\lambda) =: S_l(\lambda)$ will coincide with the corresponding market spread S_l^* for each tranche l. To be more specific, the parameters $\lambda = (\lambda_{I_j}^{(k)})_{j,k}$ are obtained according to

$$(20) \qquad \lambda = \underset{\widehat{\lambda}}{\operatorname{argmin}} \sum_l \left(\frac{S_l(\widehat{\lambda}) - S_l^*}{S_l^*} \right)^2$$

under the constraints that all elements in λ are nonnegative and that λ satisfies the inequalities (19) for every group I_l and in each time interval $[T_{k-1}, T_k]$ where $T_0 = 0$. In $S_l(\widehat{\lambda})$ we have emphasized that the model spread for tranche l is a function of $\lambda = (\lambda_{I_j}^{(k)})_{j,k}$ but we suppressed the dependence in other parameters like interest rate, payment frequency or λ_i, $i = 1, \ldots, n$.

4.2 Calibration Results

In all the numerical calibrations below we use an interest rate of 3%, the payments in the premium leg are quarterly and the integral in the default leg is discretized on a quarterly mesh. The constant recoveries for all obligors are set equal to 40%. We use Matlab in our numerical calculations and the related objective function is minimized under the suitable constraints by using the built in optimization routine fmincon (e.g. in the setup of Subsection 4.1, minimizing the criterion (20) under the constraints given by equations on the form (19)).

In this subsection we calibrate our model against CDO tranches on the iTraxx Europe and CDX.NA.IG series with maturity of five years. We use the calibration methodologies described in Subsection 4.1.

Hence, the 125 single-name CDSs constituting the entities in these series are bootstrapped from their market spreads for $T_1 = 3$ and $T_2 = 5$ using piecewise

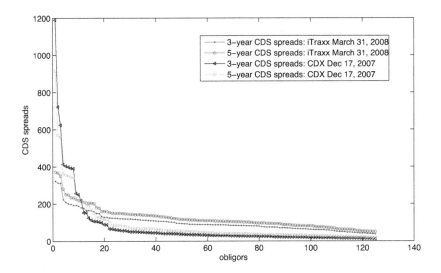

Figure 1. The 3 and 5-year market CDS spreads for the 125 obligors used in the single-name boot-strapping, for the two portfolios CDX.NA.IG sampled on December 17, 2007 and the iTraxx Europe series sampled on March 31, 2008. The CDS spreads are sorted in decreasing order.

Table 1 CDX.NA.IG Series 9, December 17, 2007 and iTraxx Europe Series 9, March 31, 2008. The market and model spreads and the corresponding absolute errors, both in bp and in percent of the market spread. The [0, 3] spread is quoted in %. All maturities are for five years.

CDX 2007-12-17

Tranche	[0, 3]	[3, 7]	[7, 10]	[10, 15]	[15, 30]
Market spread	48.07	254.0	124.0	61.00	41.00
Model spread	48.07	254.0	124.0	61.00	38.94
Absolute error in bp	0.010	0.000	0.000	0.000	2.061
Relative error in %	0.0001	0.000	0.000	0.000	5.027

iTraxx Europe 2008-03-31

Tranche	[0, 3]	[3, 6]	[6, 9]	[9, 12]	[12, 22]
Market spread	40.15	479.5	309.5	215.1	109.4
Model spread	41.68	429.7	309.4	215.1	103.7
Absolute error in bp	153.1	49.81	0.0441	0.0331	5.711
Relative error in %	3.812	10.39	0.0142	0.0154	5.218

constant individual default intensities on the time intervals [0, 3] and [3, 5]. Figure 1 displays the 3 and 5-year market CDS spreads for the 125 obligors used in the single-name bootstrapping, for the two portfolios CDX.NA.IG sampled on

December 17, 2007 and the iTraxx Europe series sampled on March 31, 2008. The CDS spreads are sorted in decreasing order.

When calibrating the joint default intensities $\lambda = (\lambda_{I_j}^{(k)})_{j,k}$ for the CDX.NA.IG Series 9, December 17, 2007 we used 5 groups $I_1, I_2, \ldots, I_5$ where $I_j = \{1, \ldots, i_j\}$ for $i_j = 6, 19, 25, 61, 125$. Recall that we label the obligors by decreasing level of riskiness. We use the average over 3-year and 5-year CDS spreads as a measure of riskiness. Consequently, obligor 1 has the highest average CDS spread while company 125 has the lowest average CDS spread. Moreover, the obligors in the set $I_5 \setminus I_4$ consisting of the 64 safest companies are assumed to never default individually, and the corresponding CDSs are excluded from the calibration, which in turn relaxes the constraints for λ in (19). Hence, the obligors in $I_5 \setminus I_4$ can only bankrupt due to a simultaneous default of the companies in the group $I_5 = \{1, \ldots, 125\}$, i.e., in an Armageddon event. With this structure the calibration against the December 17, 2007 data-set is very good as can be seen in Table 1. By using stochastic recoveries specified as in [5] one can get a perfect fit of the same data-set, see in [5]. The calibrated common shock intensities λ for the 5 groups in the December 17, 2007 data-set are displayed in the left subplot of Figure 2.

The calibration of the joint default intensities $\lambda = (\lambda_{I_j}^{(k)})_{j,k}$ for the data sampled at March 31, 2008 is more demanding. This time we use 18 groups $I_1, I_2, \ldots, I_{18}$ where $I_j = \{1, \ldots, i_j\}$ for $i_j = 1, 2, \ldots, 11, 13, 14, 15, 19, 25, 79, 125$. In order to improve the fit, as in the 2007-case, we relax the constraints for λ in (19) by excluding from the calibration the CDSs corresponding to the obligors in $I_{18} \setminus I_{17}$. Hence, we assume that the obligors in $I_{18} \setminus I_{17}$ never default individually, but can only bankrupt due to an simultaneous default of all companies in the group $I_{18} = \{1, \ldots, 125\}$. In this setting, the calibration of the 2008 data-set with constant recoveries yields an acceptable fit except for the [3, 6] tranche, as can be seen in Table 1. However, by including stochastic recoveries specified as in [5] the fit can be substantially improved, see in [5]. Furthermore, the more groups added the better the fit, which explain why we use as many as 18 groups (this holds both for constant and stochastic recoveries). The calibrated common shock intensities λ for the 18 groups in the March 2008 data-set are displayed in the right subplot of Figure 2.

Let us finally discuss the choice of the groupings $I_1 \subset I_2 \subset \ldots \subset I_m$ in our calibrations. First, for the CDX.NA.IG Series 9, December 17, 2007 data set, we used $m = 5$ groups with as always $i_m = n$. For $j = 1, 2$ and 4 the choice of i_j corresponds to the number of defaults needed for the loss process with constant recovery of 40% to reach the j-th attachment points. Hence, $i_j \cdot \frac{1-R}{n}$ with $R = 40\%$ and $n = 125$ then approximates the attachment points 3%, 10%, 30% which explains the choice $i_1 = 6, i_2 = 19, i_4 = 61$. The choice of $i_3 = 25$ implies a loss of 12% and gave a better fit than choosing i_3 to exactly match 15%. Finally, no group was chosen to match the attachment point of 7% since this made the

Table 2 Three different groupings (denoted A,B and C) consisting of $m = 7, 9, 13$ groups having the structure $I_1 \subset I_2 \subset \ldots \subset I_m$ where $I_j = \left\{1, \ldots, i_j\right\}$ for $i_j \in \{1, 2, \ldots, m\}$ and $i_1 < \ldots < i_m = 125$.

					Three different groupings								
i_j	i_1	i_2	i_3	i_4	i_5	i_6	i_7	i_8	i_9	i_{10}	i_{11}	i_{12}	i_{13}
Grouping A	6	14	15	19	25	79	125						
Grouping B	2	4	6	14	15	19	25	79	125				
Grouping C	2	4	6	8	9	10	11	14	15	19	25	79	125

Table 3 The relative calibration error in percent of the market spread, for the three different groupings A, B and C in Table 2, when calibrated against CDO tranche on iTraxx Europe Series 9, March 31, 2008 (see also in Table 1).

		Relative calibration error in % (constant recovery)			
Tranche	$[0, 3]$	$[3, 6]$	$[6, 9]$	$[9, 12]$	$[12, 22]$
Error for grouping A	6.875	18.33	0.0606	0.0235	4.8411
Error for grouping B	6.622	16.05	0.0499	0.0206	5.5676
Error for grouping C	4.107	11.76	0.0458	0.0319	3.3076

calibration worse off for all groupings we tried. With the above grouping structure we got almost perfect fits in the constant recovery case as was seen in Table 1 (and a perfect fit with stochastic recovery, see in [5]). Unfortunately, using the

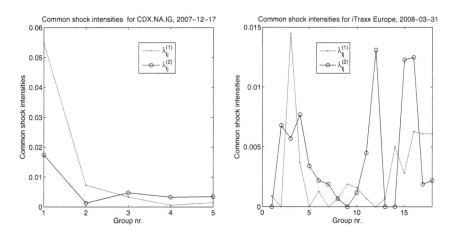

Figure 2. The calibrated common shock intensities $(\lambda_{I_j}^{(k)})_{j,k}$ for the two portfolios CDX.NA.IG sampled on December 17, 2007 (left) and the iTraxx Europe series sampled on March 31, 2008 (right).

same technique on the market CDO data from the iTraxx Europe series sampled on March 31, 2008 was not enough to achieve good calibrations. Instead more groups had to be added and we tried different groupings which led to the optimal choice rendering the calibration in Table 1. To this end, it is of interest to study the sensitivity of the calibrations with respect to the choice of the groupings on the form $I_1 \subset I_2 \subset \ldots \subset I_m$ where $I_j = \{1, \ldots, i_j\}$ for $i_j \in \{1, 2, \ldots, m\}$ and $i_1 < \ldots < i_m = 125$ on the March 31, 2008, data set. Three such groupings are displayed in Table 2 and the corresponding calibration results on the 2008 data set is showed in Table 3.

From Table 3 we see that the relative calibration error in percent of the market spread decreased monotonically for the first three thranches as the number of groups increased. The rest of the parameters in the calibration where the same as in the optimal calibration in Table 1.

Finally, we remark that the two optimal groupings used in Table 1 in the two different data sets CDX.NA.IG Series 9, December 17, 2007 and iTraxx Europe Series 9, March 31, 2008 differ quite a lot. However, the CDX.NA.IG Series is composed by North American obligors while the iTraxx Europe Series is formed by European companies. Thus, there is no model risk or inconsistency created by using different groupings for these two different data sets, coming from two disjoint markets. If on the other hand the same series is calibrated and assessed (e.g. for hedging) at different time points in a short time span, it is of course desirable to use the same grouping in order to avoid model risk.

4.2.1 The Implied Loss Distribution

After the fit of the model against market spreads we can use the calibrated portfolio parameters $\lambda = (\lambda_{I_j}^{(k)})_{j,k}$ together with the calibrated individual default intensities, to study the credit-loss distribution in the portfolio. In this paper we only focus on some few examples derived from the loss distribution with constant recoveries evaluated at $T = 5$ years.

The allowance of joint defaults of the obligors in the groups I_j together with the restriction of the most safest obligors not being able to default individually, will lead to some interesting effects of the loss distribution, as can be seen in Figures 3 and 4. For example, we clearly see that the support of the loss-distributions will in practice be limited to a rather compact set. To be more specific, the graphs in Figure 3 indicate that $\mathbb{P}[N_5 = k]$ roughly has support on the set $\{1, \ldots, 35\} \cup \{61\} \cup \{125\}$ for the 2007 case and on $\{1, \ldots, 40\} \cup \{79\} \cup \{125\}$ for the 2008 data-set. This becomes even more clear in a log-loss distribution, as is seen in Figure 4.

From the left graph in Figure 4 we see that the default-distribution is nonzero on $\{36, \ldots, 61\}$ in the 2007-case and nonzero on $\{41, \ldots, 79\}$ for the 2008-sample, but the actual size of the loss-probabilities are in the range 10^{-10} to 10^{-70}. Such low values will obviously be treated as zero in any practically relevant computa-

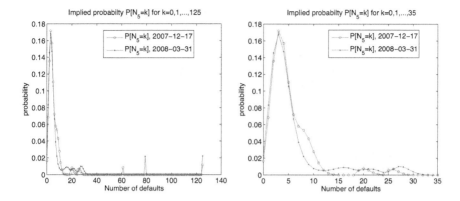

Figure 3. The implied distribution $\mathbb{P}[N_5 = k]$ on $\{0, 1, \ldots, \ell\}$ where $\ell = 125$ (left) and $\ell = 35$ (right) when the model is calibrated against CDX.NA.IG Series 9, December 17, 2007 and iTraxx Europe Series 9, March 31, 2008.

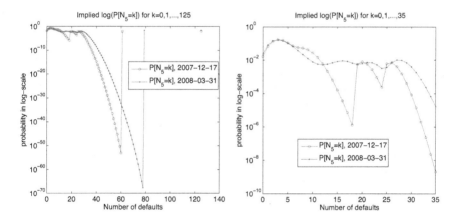

Figure 4. The implied log distribution $\ln(\mathbb{P}[N_5 = k])$ on $\{0, 1, \ldots, \ell\}$ where $\ell = 125$ (left) and $\ell = 35$ (right) when the model is calibrated against CDX.NA.IG Series 9, December 17, 2007 and iTraxx Europe Series 9, March 31, 2008.

tion. Furthermore, the reasons for the empty gap in the left graph in Figure 4 on the interval $\{62, \ldots, 124\}$ for the 2007-case is due to the fact that we forced the obligors in the set $I_5 \setminus I_4$ to never default individually, but only due to an simultaneous common shock default of the companies in the group $I_5 = \{1, \ldots, 125\}$. This Armageddon event is displayed as an isolated nonzero 'dot' at default nr 125 in the left graph of Figure 4. The gap on $\{80, \ldots, 124\}$ in the 2008 case is explained similarly due to our assumption on the companies in the set $I_{19} \setminus I_{18}$. Also note that the two 'dots' at default nr 125 in the left subplot of Figure 4 are manifested as spikes in the left graph displayed in Figure 3. The shape of the multimodal loss distributions presented in Figure 3 and Figure 4 are typical for models allowing simultaneous defaults, see for example Figure 2, page 59 in [11] and Figure 2, page 710 in [14].

5. Min-Variance Hedging

In this section we present some numerical results illustrating performance of the min-variance hedging strategies given in Proposition 3.2 of [4]. This will be done in the setup of the calibrated model of Subsection 4.1 (model calibrated with constant recoveries to the CDX.NA.IG Series 9 data set of December 17, 2007).

The aim of this subsection is to analyze the composition of the hedging portfolio at time $t = 0$ (the calibration date) when standardized CDO tranches are hedged with a group of d single-name CDSs, which are included in the underlying CDS index. Since no spread factor $\mathbf{X}$ is used in the model, Proposition 3.2 of [4] then implies that the min-variance hedging ratios at time $t = 0$ is given by $\zeta^{va}(0, \mathbf{H}_0) = (u, \mathbf{v})(\mathbf{v}, \mathbf{v})^{-1}(0, \mathbf{H}_0)$ where

$$(u, \mathbf{v}) = \sum_{Y \in \mathcal{Y}} \lambda_Y(0) \Delta u^Y (\Delta \mathbf{v}^Y)^\mathsf{T} \quad \text{and} \quad (\mathbf{v}, \mathbf{v}) = \sum_{Y \in \mathcal{Y}} \lambda_Y(0) \Delta \mathbf{v}^Y (\Delta \mathbf{v}^Y)^\mathsf{T}.$$

Hence, computing the min-variance hedging ratios involves a summation of the "jump differentials" $\lambda_Y(0) \Delta u^Y (\Delta \mathbf{v}^Y)^\mathsf{T}$ and $\lambda_Y(0) \Delta \mathbf{v}^Y (\Delta \mathbf{v}^Y)^\mathsf{T}$ over all possible triggering events $Y \in \mathcal{Y}$ where $\mathcal{Y} = \{\{1\}, \ldots, \{n\}, I_1, \ldots, I_m\}$.

In the calibration of the CDX.NA.IG Series 9, we used $m = 5$ groups $I_1, I_2, \ldots, I_5$ where $I_j = \{1, \ldots, i_j\}$ for $i_j = 6, 19, 25, 61, 125$ and the obligors have been labeled by decreasing level of riskiness. At the calibration date $t = 0$ associated with December 17, 2007, no name has defaulted in CDX Series 9 so we set $\mathbf{H}_0 = \mathbf{0}$. In our empirical framework, the intensities $\lambda_Y(0)$, $Y \in \mathcal{Y}$ are computed from the constant default intensities λ_i that fit market spreads of 3-year maturity CDSs and from the 3-year horizon joint default intensities λ_{I_j} calibrated to CDO tranche quotes. The terms $\Delta u^Y(0, \mathbf{H}_0)$ and $\Delta \mathbf{v}^Y(0, \mathbf{H}_0)$ corresponds to the change in value of the tranche and the single-name CDSs, at the arrival of the triggering event affecting all names in group Y. Recall that the cumulative change in value of the tranche is equal to

$$(21) \qquad \Delta u^Y(0, \mathbf{H}_0) = L_{a,b}(\mathbf{H}_0^Y) - L_{a,b}(\mathbf{H}_0) + u(0, \mathbf{H}_0^Y) - u(0, \mathbf{H}_0)$$

where $\mathbf{H}_0^Y$ is the vector of $\{0, 1\}^n$ such that only the components $i \in Y$ are equal to one. Hence, the tranche sensitivity $\Delta u^Y(0, \mathbf{H}_0)$ includes both the protection payment on the tranche associated with the default of group Y and the change in the ex-dividend price u of the tranche. Note that the price sensitivity is obtained by computing the change in the present value of the default leg and the premium leg.

Table 4 The names and CDS spreads (in bp) of the six riskiest obligors used in the hedging strategy displayed by Figure 5.

Company (Ticker)	CCR-HomeLoans	RDN	LEN	SFI	PHM	CTX
3-year CDS spread	1190	723	624	414	404	393

The latter quantity involves the contractual spread that defines cash-flows on the premium leg. As for CDX.NA.IG Series 9, the contractual spreads were equal to 500 bps, 130 bps, 45 bps, 25 bps and 15 bps for the tranches [0-3%], [3-7%], [7-10%], [10-15%] and [15-30%]. We use the common-shock interpretation to compute $u(0, \mathbf{H}_0^Y)$ and $u(0, \mathbf{H}_0)$ with the convolution recursion pricing scheme detailed in Subsection 3. More precisely, using the same notation as in Subsection 3, the CDO tranche price $u(0, \mathbf{H}_0^Y)$ (resp. $u(0, \mathbf{H}_0)$) is computed using the recursion procedure with $Z = N_n \setminus Y$ (resp. $Z = N_n$). We let $i_1, \ldots i_d$ be the CDSs used in the min-variance hedging and assume that they all are initiated at time $t = 0$. Hence, the market value at $t = 0$ for these CDSs are zero. As a result, when group Y defaults simultaneously, the change in value $\Delta \mathbf{v}^Y(0, \mathbf{H}_0)$ for buy-protection positions on these CDSs is only due to protection payment associated with names in group Y. Hence, for one unit of nominal exposure on hedging CDSs, the corresponding vector of sensitivities is equal to $\Delta \mathbf{v}^Y(0, \mathbf{H}_0) = ((1 - R)\mathbb{1}_{i_1 \in Y}, \ldots, (1 - R)\mathbb{1}_{i_d \in Y})^\top$ where the recovery rate R is assumed to be constant and equal to 40%.

Figure 5 displays the nominal exposure for the d most riskiest CDSs when hedging one unit of nominal exposure in a CDO by using the min-variance hedging strategy in Proposition 3.2 of [4]. We use $d = 3, 4, 5$ and $d = 6$ in our computations. Furthermore, Table 4 displays the names and sizes of the 3-year CDS spreads used in the hedging strategy. Each plot in Figure 5 should be interpreted as follows: in every pair (x, y) the x-component represents the size of the 3-year CDS spread at the hedging time $t = 0$ while the y-component is the corresponding nominal CDS-exposure computed via Proposition 3.2 of [4] using the d riskiest CDSs. The graphs are ordered from top to bottom, where the top panel corresponds to hedging with the $d = 3$ riskiest CDS and the bottom panel corresponds to hedging with the $d = 6$ riskiest names. Note that the x-axes are displayed from the riskiest obligor to the safest. Thus, hedge-sizes y for riskier CDSs are aligned to the left in each plot while y-values for safer CDSs are consequently displayed more to the right. In doing this, going from the top to the bottom panel consists in observing the effect of including new safer names from the right part of the graphs. We have connected the pairs (x, y) with lines forming graphs that visualizes possible trends of the min-variance hedging strategies for the d most riskiest CDSs.

For example, when the three riskiest names are used for hedging (top panel), we observe that the amount of nominal exposure in hedging instruments decreases

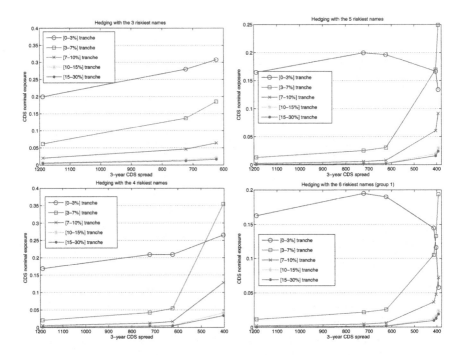

Figure 5. Min-variance hedging strategies associated with the d riskiest CDSs, $d = 3, 4, 5, 6$ for one unit of nominal exposure of different CDO tranches in a model calibrated to market spreads of CDX.NA.IG Series 9 on December 17, 2007.

with the degree of subordination, i.e., the [0-3%] equity tranche requires more nominal exposure in CDSs than the upper tranches. Note moreover that the min-variance hedging portfolio contains more CDSs on names with lower spreads. When lower-spread CDSs are added in the portfolio, the picture remains almost the same for the 3 riskiest names. For the remaining safer names however, the picture depends on the characteristics of the tranche. For the [0-3%] equity tranche, the quantity of the remaining CDSs required for hedging sharply decrease as additional safer names are added. One possible explanation is that adding too many names in the hedging strategy will be useless when hedging the equity tranche. This is intuitively clear since one expects that the most riskiest obligors will default first and consequently reduce the equity tranche substantially, explaining the higher hedge-ratios for riskier names, while it is less likely that the more safer names will default first and thus incur losses on the first tranche which explains the lower hedge ratios for the safer names. We observe the opposite trend for the senior (safer) tranches: adding new (safer) names in the hedging portfolio seems

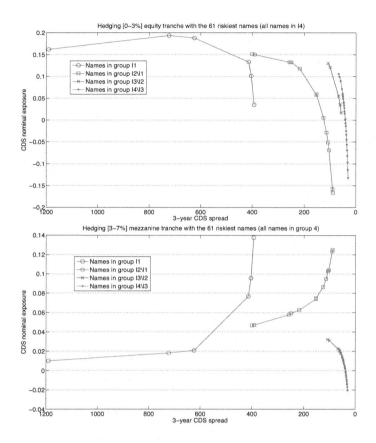

Figure 6. Min-variance hedging strategies when hedging one unit of nominal exposure in the [0-3%] equity tranche (top) and the [3-7%] mezzanine tranche (bottom) using the d riskiest CDSs, $d = 61$ (all names excepted names in group $I_5 \setminus I_4$) for one unit of nominal exposure.

to be useful for "non equity" tranches since the nominal exposure required for these names increases when they are successively added.

Figure 6 and 7 display min-variance hedging strategies when hedging a standard tranche with the 61 riskiest names, i.e., all names excepted names in group $I_5 \setminus I_4$. Contrary to Figure 5, these graphs allow to visualize the effect of the "grouping structure" on the composition of the hedging portfolio. In this respect, we use different marker styles in order to distinguish names in the different disjoint groups I_1, $I_2 \setminus I_1$, $I_3 \setminus I_2$, $I_4 \setminus I_3$. As one can see, the min-variance hedging strategies are quite different among tranches. Moreover, whereas nominal exposures required for hedging are monotone for names belonging to the same disjoint group (except group I_1 for the equity tranche), this tendency is broken when we

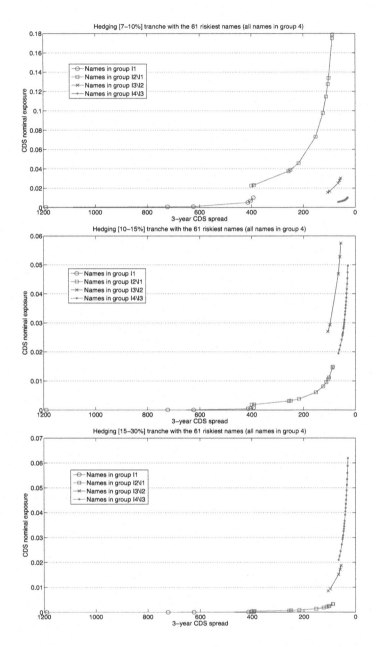

Figure 7. Min-variance hedging strategies when hedging one unit of nominal exposure in the [7-10%] tranche (top), the [10-15%] tranche (middle) and the [15-30%] tranche (bottom) with the d riskiest CDSs, $d = 61$ (all names excepted names in group $I_5 \setminus I_4$).

consider names in different groups. This suggests that the grouping structure has a substantial impact on the distribution of names in the hedging portfolio.

For the equity tranche, we observe in Figure 5 that less safer-names are required for hedging. This feature is retained in Figure 6 when we look at names in specific disjoint groups. Indeed, names in a given disjoint group are affected by the same common-shocks which in turn affect the equity tranche with the same severity. The only effect that may explain differences in nominal exposure among names in the same disjoint group is spontaneous defaults: names with wider spreads are more likely to default first, then we expect them in greater quantity for hedging the equity tranche than names with tighter spreads. This intuition is not true for all names since the nominal exposure for the riskiest name is lower than for the second riskiest name. This is due to the fact that the change in value for the equity tranche in (21) is greater when the second riskiest name default spontaneously.

Note that nominal exposure in hedging CDS even becomes negative for names within groups $I_2 \setminus I_1$ and $I_4 \setminus I_3$ when spreads are low. However, in Figure 6 we observe that, for the equity tranche, some of the riskiest names in $I_4 \setminus I_3$ are more useful in the hedging than some of the safest names in group I_1, which may sound strange at a first glance, given that the credit spread of the latter is much larger than the credit spread of the former. Recall that the equity tranche triggers protection payments corresponding to the few first defaults, if these occur before maturity. Even if names in group $I_4 \setminus I_3$ have a very low default probability, the fact that they can affect the tranche at the arrival of common-shocks I_4 or I_5 makes these names appealing for hedging because they are less costly (they require less premium payments) than names in I_1.

Figure 6 suggests that names with the lowest spreads should be ineffective to hedge the [0-3%] and the [3-7%] tranches. As can be seen in Figure 7, this is the contrary for the other tranches, i.e., the amount of low-spread names in the hedging portfolio increases as the tranche becomes less and less risky. For the [15-30%] super-senior tranche, we can see on the lowest graph of Figure 7 that the safer a name is, the larger the quantity which is required for hedging. Furthermore, Figure 7 also shows that in a consistent dynamic model of portfolio credit risk calibrated to a real data set, the [15-30%] super-senior tranche has significant (in fact, most of its) sensitivity to very safe names with spreads less than a few dozens of bp-s. For this tranche it is actually likely that one could improve the hedge by inclusion of even safer names to the set of hedging instruments, provided these additional names could also be calibrated to. Recall that on the data of CDX.NA.IG Series 9 on December 17, 2007, we calibrated our model to the 64 safest names in the portfolio.

References

1. Andersen, L. and Sidenius, J.: Extensions to the Gaussian Copula: Random Recovery

and Random Factor Loadings, *Journal of Credit Risk*, Vol. 1, No. 1 (Winter 2004), p. 29–70.

2. Assefa, S., Bielecki, T.R., Crépey, S. and Jeanblanc, M.: CVA computation for counterparty risk assessment in credit portfolios. *Credit Risk Frontiers*, Bielecki, T.R., Brigo, D. and Patras, F., eds., Wiley/Bloomberg-Press, 2011.

3. Bielecki, T.R., Cousin, A., Crépey, S., Herbertsson, A.: Dynamic Hedging of Portfolio Credit Risk in a Markov Copula Model. Forthcoming in *Journal of Optimization Theory and Applications*.

4. Bielecki, T.R., Cousin, A., Crépey, S., Herbertsson, A.: A Bottom-Up Dynamic Model of Portfolio Credit Risk – Part I: Markov copula perspective. In *Recent Advances in Financial Engineering 2012*, World Scientific, 2013.

5. Bielecki, T.R., Cousin, A., Crépey, S., Herbertsson, A.: A bottom-up dynamic model of portfolio credit risk with stochastic intensities and random recoveries, forthcoming in *Communications in Statistics—Theory and Methods*.

6. Bielecki, T. R. and Crépey, S.: Dynamic Hedging of Counterparty Exposure. Forthcoming in *The Musiela Festschrift*, Zariphopoulou, T., Rutkowski, M. and Kabanov, Y., eds, Springer.

7. Bielecki, T.R., Crépey, S., Jeanblanc, M.: Up and down credit risk. *Quantitative Finance* 10 (10), pp. 1137–1151 (2010).

8. Bielecki, T.R., Crépey, S., Jeanblanc, M. and Zargari, B.: Valuation and Hedging of CDS Counterparty Exposure in a Markov Copula Model. *International Journal of Theoretical and Applied Finance* 15 (1) 1250004, 2012.

9. Bielecki, T. R. and Jakubowski, J. and Niewęglowski, M.: Dynamic Modeling of Dependence in Finance via Copulae Between Stochastic Processes, *Copula Theory and Its Applications*, Lecture Notes in Statistics, Vol.198, Part 1, 33–76, 2010.

10. Brigo, D., Pallavicini, A., Torresetti, R. Cluster-based extension of the generalized poisson loss dynamics and consistency with single names. *International Journal of Theoretical and Applied Finance*, Vol 10, n. 4, 607-632, 2007.

11. Brigo, D., Pallavicini, A., Torresetti, R. Credit models and the crisis: default cluster dynamics and the generalized Poisson loss model, *Journal of Credit Risk*, 6 (4), 39–81, 2010.

12. S. Crépey, T. R. Bielecki and D. Brigo (2013): *Counterparty Risk and Funding*, Taylor & Francis (in preparation).

13. Crépey, S. and Rahal, A.: Simulation/Regression Pricing Schemes for CVA Computations on CDO Tranches, forthcoming in *Communications in Statistics—Theory and Methods*.

14. Elouerkhaoui, Y.: Pricing and Hedging in a Dynamic Credit Model. *International Journal of Theoretical and Applied Finance*, Vol. 10, Issue 4, 703–731, 2007.

15. Iscoe, I., Jackson, K., Kreinin, A. and Ma, X.: On Exponential Approximation to the Hockey Stick Function. *Working Paper, Department of Computer Science, University of Toronto*, 2010.

16. Iscoe, I., Jackson, K., Kreinin, A. and Ma, X.: Pricing correlation-dependent derivatives based on exponential approximations to the hockey stick function (working paper), 2007.

17. Beylkin, G. and Monzon, L. On approximation of functions by exponential sums, *Applied and Computational Harmonic Analysis*, 19 (1): 17-48, 2005.

18. Marshall, A. & Olkin, I.: A multivariate exponential distribution, *J. Amer. Statist. Assoc.*, 2, 84-98, 1967.
19. Patton, A.: Modelling Time-varying exchange rate dependence using the conditional copula, *Working Paper* 2001-09, University of California, San Diego, 2001.

On the Limit Behavior of Option Hedging Sets under Transaction Costs

J. Grépat

Laboratoire de Mathématiques, Université de Franche-Comté, 16 Route de Gray, 25030
Besançon, cedex, France
Email: julien.grepat@univ-fcomte.fr

In this note we link the Kusuoka limit theorem on super replication of
European contingent claim under transaction costs with the multi–asset
mainstream and we study the asymptotic behavior of the hedging sets in
the context of topological convergence of subsets of $\mathbb{R}^2$.

Key words: transaction costs, Kusuoka theorem, super replication

1. Introduction

With the rising of financial mathematics and the growing success of
continuous–time models, more and more sophisticated theories are developed
though the real finance world is still discrete. Even if time steps decrease as much
as technologies permit it, the orders, actualizations of price are made along a discrete time grid. The question of linking discrete– and continuous–time models is
laid and paradoxes appears. Indeed, it is well-known that the straightforward discrete approximation of a continuous time model may not lead to the convergence
of the option price. In [9], the remedy is the introduction of decreasing transaction
costs, see also book [6], references therein and more recent papers [1], [10]. This
point of view is logical with the paradigm of transaction costs. The agents agree
on a certain level of the latter given a number of actualizations. For a growing
number of revisions, the transaction costs coefficients will naturally decrease. In
the Leland–Lott models, the terminal values of portfolios approximate the pay-off
of the option and the limit of their initial values is declared to be a fair option price
accepted by practitioners as realistic.

In Kusuoka [8], it is considered a sequence of two-asset models where the
proportional transaction costs tend to zero with rate $n^{-1/2}$ and it is calculated the
asymptotic replication price of a rather general European option. It turns out that
this price is different from that of the limiting continuous–time model based on a
geometric Brownian motion and has to be calculated over an enlarged family of

continuous martingales.

The aim of this paper is to set the Kusuoka approach in the framework of the multi–asset generalization of [2]. In the multi–asset theory, developed in [6], we consider a vector–valued price process where each coordinate represents the price evolution of a risky asset in quotes of a numéraire. All usual objects, portfolios, contingent claims, initial endowments, etc. are d-dimensional vector–valued and each component is the investment in the corresponding asset. Analogously, everything can be defined in term of physical units. Accordingly, the initial endowments from which it is possible to super replicate the European option is a subset of $\mathbb{R}^d$. These sets are characterized by the hedging theorem, a fundamental result in which a key role is played by the set of consistent price systems, that is martingales evolving in the dual of the solvency region. "Vectorization" of the theory fills the gap between the approach of classical mathematical finance (where everything is expressed in money) and that of mathematical economics (where the vectors of commodities can be considered as primary objects).

We shall detail with care the links between the traditional argument used in [8] with the geometric approach of [2], restricted to two-asset models. Our models are essentially the same as that of Kusuoka, [8]. The minor differences are in the use of structures, reasoning and notations compatible with the now standard theories described in [5] and [6]. In [8] the subject of interest is the limiting behavior of the points x_n laying in the intersection of the boundary of hedging sets Γ^n with the abscissae axis, that is the minimal initial endowment in money (with a zero position in stock) needed to hedge the option. In this paper, we study the limiting behavior of the whole sets in the closed convergence topology. This paper is the first step to the generalization to the multi–asset models of [2] and gives also arguments to consider non–symmetric transaction costs therein.

2. Notations

We use the following notations:

- for a vector $v = (v^1, v^2) \in \mathbb{R}^2$,

$$|v| := \max\{|v^1|; |v^2|\};$$

- for vectors $v, w \in \mathbb{R}^2$,

$$vw = v^1 w^1 + v^2 w^2;$$

- the notations e^1, e^2 stand for the canonical vectors $(1, 0)$ and $(0, 1)$, $\mathbf{1}$ is the vector $(1, 1)$;

- the cone spanned by the family of vectors $\{v_i;\ i \in I\}$, denoted by cone $\{v_i;\ i \in I\}$, is the set of vectors

$$w = \sum_{i \in I} \alpha_i v_i; \qquad \alpha_i \in \mathbb{R}_+;$$

- for a sequence of random variables $(\xi^n)_{n\in\mathbb{N}}$ the symbol $O(n^{-a})$ means that there exists a positive constant κ such that $n^a|\xi^n| \le \kappa$ a.s. for all n;

- the symbol $H \cdot S$ stands for the stochastic integral of H with respect to S;

- $\mathbb{D}(\mathbb{R}^d)$ is the Skorohod space of the cádlág functions $x : [0, T] \to \mathbb{R}^d$ while $\mathbb{C}(\mathbb{R}^d)$ denotes the space of continuous functions taking values in $\mathbb{R}^d$ with the uniform norm

$$\|x\|_T = \sup_{t\le T} |x_t|.$$

For a survey of Skorohod topology, weak convergence in the Skorohod space and corresponding notations we refer to the book [5].

3. Model and main result

We consider 2-asset models of currency market with transaction costs following the ideas of the book [6]. The first non-risky asset will serve as the numéraire, the second is risky. An asset can be exchanged to the other paying the proportional transaction costs. That is to increase the value of the jth position in one unit (of numéraire), one need to diminish in $(1 + \lambda^{ij})$ unit (of numéraire) the ith position. Namely, the models are given by transaction costs matrices. We fix as basic parameter the 2-square matrix Λ with zero diagonal and positive entries. We consider the transaction cost matrix for the n-th model

$$\Lambda^n = \Lambda \sqrt{T/n}.$$

Price processes

We define in this subsection continuous-time models whose price processes are piecewise constant on the intervals forming uniform partitions of $[0, T]$. Of course, these models are in one-to-one correspondence with discrete–time models. Fix the drift and volatility parameters $\mu \in \mathbb{R}$, $\sigma \in]0, \infty[$ and put, for $n \ge 1$,

$$\mu^n = \mu T/n, \qquad \sigma^n = \sigma \sqrt{T/n}.$$

On the probability space $(\Omega, \mathcal{F}, P)$, we consider, for each n, a family of i.i.d. random variables $\{\xi_k; \ k \le n\}$, where ξ_k take values in $\{-1, 1\}$ and $P(\xi_k = 1) = 1/2$. Put

$$t_k = t_k^n := kT/n.$$

The process S^{n2} models the price evolution of one unit of the risky security measured in units of the first non-risky asset serving as numéraire. We define the process $S_t^n = (S_t^{n1}, S_t^{n2})$ where $S_t^{n1} = 1$ and

$$S_0^{n2} = 1, \qquad S_t^{n2} = \prod_{m=1}^{k} (1 + \mu^n + \sigma^n \xi_m), \qquad t \in [t_k, t_{k+1}[,$$

for sufficiently large n (to insure that $S^{n2} > 0$). In this setting the stochastic basis is $(\Omega, \mathcal{F}, \mathbf{F^n}, P)$ where the filtration $\mathbf{F^n} = (\mathcal{F}_t^n)$ is $\mathcal{F}_t^n := \sigma\{S_r^n, \ r \leq t\}$.

Transaction costs

The solvency region is the cone defined by

$$K^{\Lambda^n} = \text{cone}\left\{\left(1 + \lambda^{n12}\right)e^1 - e^2, \left(1 + \lambda^{n21}\right)e^2 - e^1\right\},$$

that is K^{Λ^n} is the set of positions which can be converted, paying transaction costs, to get only non-negative amount on each asset. The (positive) dual cone is the set

$$K^{\Lambda^n*} = \left\{w \in \mathbb{R}^2 : \frac{1}{1 + \lambda^{n21}} \leq \frac{w^2}{w^1} \leq 1 + \lambda^{n12}\right\},$$

which is the set of vectors with a non-negative scalar product with any vector of K^{Λ^n}.

The piecewise constant process V solving the linear controlled stochastic equation

$$V_0 = v \in K^{\Lambda^n}, \quad dV_t^i = V_{t-}^i dS_t^{ni}/S_{t-}^{ni} + dB_t^i, \quad i = 1, 2,$$

models the portfolio value process with strategy B, where the components of the control B are

$$B^i = \sum_{k=1}^n B_k^i \mathbb{I}_{]t_{k-1}, t_k]},$$

B_k^i is $\mathcal{F}_{t_{k-1}}^n$-measurable and $\Delta B_{t_k} = B_{t_k} - B_{t_{k-1}} \in L^0(-K^{\Lambda^n}, \mathcal{F}_{t_{k-1}}^n)$. The set of such processes V with initial value v is denoted by $\mathcal{A}_v^n$ while the notation $\mathcal{A}_v^n(T)$ is reserved for the set of their terminal value V_T.

Using the random diagonal operator

$$\phi_t^n : (x^1, x^2) \mapsto (x^1, x^2/S_t^{n2})$$

define the random cone $\widehat{K}_t^{\Lambda^n} = \phi_t^n K^{\Lambda^n}$ with the dual $\widehat{K}_t^{\Lambda^n*} = (\phi_t^n)^{-1} K^{\Lambda^n*}$.

Hedging sets

Our aim is to price a European option. We shall consider a two-dimensional pay-off. The first asset is an amount of money in numéraire, whereas the second is a quantity of physical units. The pay-off is of the form $F(S^n)$ with the function $F : \mathbb{D}(\mathbb{R}^2) \to \mathbb{R}_+^2$ supposed to be bounded and continuous in the Skorohod topology on $\mathbb{D}(\mathbb{R}^2)$. Let Γ^n be the set of initial endowments from which one can start a self-financing portfolio process with the terminal value dominating the contingent claim $F(S^n)$, that is

$$\Gamma^n = \{v \in \mathbb{R}^2 : (\phi_T^n)^{-1} F(S^n) \in \mathcal{A}_v^n(T) \text{ a.s.}\}.$$

We denote by $\mathcal{M}^n$ the set of all $\mathbf{F}^n$-martingales Z such that $Z_t \in \widehat{K}_t^{\wedge n*} \setminus \{0\}$ a.s. and $Z_0^1 = 1$. According to [6], Chap. 3,

$$(3.1) \qquad \Gamma^n = \left\{ v \in \mathbb{R}^2 : vZ_0 \geq EZ_T F(S^n) \text{ for all } Z \in \mathcal{M}^n \right\}.$$

This identity is the so-called hedging theorem claiming that one can super replicate the contingent claim if and only if the value of the initial endowments is not less than the expectation of the value of the contingent claim whatever a consistent price system is used to the comparison. The theorem holds under the assumption of the existence of a strictly consistent price system, fulfilled for our models.

Limit sets and main results

In analogy with the use of consistent price systems for the hedging theorem, we shall define the following set of martingales. Let B be a Brownian motion. We define $\mathcal{M}$ as the set of processes $(1, M)$,

$$M = \mathcal{E}(g \cdot B),$$

where g is a predictable adapted process whose square admits the following bounds:

$$\sigma(\sigma - 2\lambda) \leq g^2 \leq \sigma(\sigma + 2\lambda),$$

with λ be the mean of the transaction costs coefficients,

$$\lambda = \frac{\lambda^{12} + \lambda^{21}}{2}.$$

We put

$$\Gamma = \left\{ v \in \mathbb{R}^2 : vZ_0 \geq EZ_T F(Z) \text{ for all } Z \in \mathcal{M} \right\}.$$

The main results of this note are the following. In the formulation of Theorem 3.1 below, we could refer to convergence in the closed topology of the subsets of $\mathbb{R}^2$, see [4]. We provide a simple but equivalent characterization in terms of sequences.

Theorem 3.1. *We have the convergence results,*

1. *for any $v \in \Gamma$, there is a sequence $v^n \in \Gamma^n$, such that $v^n \to v$,*

2. *for any convergent subsequence of the sequence $v^n \in \Gamma^n$, the limit belongs to Γ.*

We also give the following auxiliary result. In [8], the value of interest in Γ^n is following:

$$x^n = \min \left\{ v^1 : v \in \Gamma^n \cap \mathbb{R}_+ e^1 \right\}.$$

This is the minimal initial capital with a zero position in the risky asset which hedge the option.

Theorem 3.2. *The sequence $\{x_n\}$ converges to x where*

$$x = \min\left\{v^1 : v \in \Gamma \cap \mathbb{R}_+ e^1\right\}.$$

4. Weak convergence

We obtain our convergence result for Γ^n by using the representation (3.1) and the theory of weak convergence of measures. In order to make argument more transparent, it is usefull to consider a family of rather simpler polyhedral conic models in the spirit of the paper [3]. Indeed, there exists a sequence of positive numbers $\kappa^n = O(n^{-1/2})$ such that $K^{\Lambda^n *} \subset K^{\kappa^n *}$, where

$$K^{\kappa *} := \mathbb{R}_+(\mathbf{1} + U_\kappa) \qquad U_\kappa := \{v \in \mathbb{R}^2 : |v| \le \kappa\}.$$

That is, $K^{\kappa *}$ is the closed convex cone in $\mathbb{R}^2$ generated by the max-norm ball of radius κ with center at $\mathbf{1}$.

Let a sequence $Z^n \in \mathcal{M}^n$. It is easily seen that Z^n takes values in the cone $(\phi^n)^{-1} K^{\kappa^n *}$. The strictly positive martingale Z^{n1} is the density process of the probability measure $Q^n = Z_T^{n1} P$ and the process $M^n := Z^{n2}/Z^{n1}$ is a strictly positive Q^n-martingale with respect to the filtration $\mathbf{F}^n$. Observe that

(4.2)
$$\frac{1 - \kappa^n}{1 + \kappa^n} S^{n2} \le M^n \le \frac{1 + \kappa^n}{1 - \kappa^n} S^{n2}.$$

We shall show that the sequence M^n is Q^n-tight.

It is worth to note that there is a one-to-one correspondence between $\mathcal{M}^n$ and the set of "preconsistent price systems" of Kusuoka [8], it is particularly clear with the proposition 2.14 therein.

Let us define the piecewise constant processes ("stochastic logarithms" of M^n)

$$L^n := (M_-^n)^{-1} \cdot M^n.$$

Note that L^n has jumps only at the points t_k,

$$\Delta L_{t_k}^n = (M_{t_k-}^n)^{-1} \Delta M_{t_k}^n = (M_{t_{k-1}}^n)^{-1}(M_{t_k}^n - M_{t_{k-1}}^n), \qquad k \ge 1.$$

Tightness

The following lemma collects the basic asymptotics needed to check the tightness of the laws $\mathcal{L}(M^n|Q^n)$ on the Skorohod space.

Lemma 4.1. *We have the following asymptotic relations:*

(4.3)
$$\|\Delta \ln M^n\|_T = O(n^{-1/2}),$$

(4.4)
$$\|\Delta L^n\|_T = O(n^{-1/2}),$$

(4.5)
$$\|\Delta \ln M^n - \Delta L^n\|_T = O(n^{-1}),$$

(4.6)
$$\sup_{k \le n} \left| E^{Q^n}[\Delta \ln M^n_{t_k} | \mathcal{F}^n_{t_{k-1}}] \right| = O(n^{-1}).$$

Proof. We derive from (4.2) the bounds

$$- 2 \ln \frac{1 + \kappa^n}{1 - \kappa^n} + \ln (1 + \mu^n - \sigma^n)$$
$$\le \Delta \ln M^n$$
$$\le 2 \ln \frac{1 + \kappa^n}{1 - \kappa^n} + \ln (1 + \mu^n + \sigma^n),$$

implying (4.3). In view of the relation

$$\Delta L^n_{t_k} = \exp(\Delta \ln M^n_{t_k}) - 1,$$

we get (4.4). Setting

$$\Phi_1(z) := \ln(1 + z) - z = O(z^2), \qquad z \to 0,$$

the asymptotic

$$\|\Phi_1(\Delta L^n)\|_T = O(n^{-1})$$

is a consequence of (4.4). Note that

$$\Delta \ln M^n_{t_k} = \Delta L^n_{t_k} + \Phi_1\left(\Delta L^n_{t_k}\right),$$

and (4.5), (4.6) follows. □

Lemma 4.2. *Let* $Z^n \in \mathcal{M}^n$, $M^n := Z^{n2}/Z^{n1}$, *and* $Q^n := Z^{n1}_T P$. *Then:*

1. *the sequence* M^n *is* Q^n-C-*tight;*

2. *the sequence* S^n *is* Q^n-*tight and*

   (4.7)
   $$\left\| S^{n2} - M^n \right\|_T \le \|M^n\|_T \, O(n^{-1/2}).$$

Proof. Following the lines of [2], Lemma 4.2, or [8], Lemma 4.8, we get bounds for the processes M^n and their bracket's oscillations. That is, for any $m > 1$, we have

(4.8) $$\sup_n E^{Q^n}\|M^n\|_T^{2m} < \infty \qquad \text{and} \qquad \sup_n E^{Q^n}\|\ln M^n\|_T^{2m} < \infty,$$

and the following estimate for the increments of quadratic characteristics:

$$(4.9) \qquad E^{Q^n} \sup_{k \le n-l} \left| \langle M^n \rangle_{t_{k+l}} - \langle M^n \rangle_{t_k} \right|^2 \le C(l/n)^2, \qquad l \le n,$$

where the constant C does not depend on l, n. The tightness of the sequence $\mathscr{L}(M^n | Q^n)$ follows, see [5]. Furthermore, we can deduce from Lemma 4.1 that the jumps tend to zero, which shows that each limit point of the sequence of laws $\mathscr{L}(M^n | Q^n)$ is continuous by virtue of Proposition VI.3.26 in [5].

From (4.2), we easily deduce (4.7) and the following,

$$\| \ln S^{n2} \|_T \le \ln \frac{1 + \kappa^n}{1 - \kappa^n} + \| \ln M^n \|_T .$$

Which proves the second assertion. □

Identification of the limit laws

In this paragraph, we show that each limit law of the sequence $\mathscr{L}(Z^n / Z^{n1} | Q^n)$ is the law of a process in $\mathcal{M}$. With the definition of the processes of $\mathcal{M}$, one can see that we need an estimation of the quadratic variation process of L^n. This is the aim of Lemma 4.3 below.

Lemma 4.3. *We have the following asymptotic relations:*

$$-2E^{Q^n} [\ln M^n_{t_{k+l}} - \ln M^n_{t_k} | \mathcal{F}^n_{t_k}] \le (l/n) T \sigma(\sigma + 2\lambda) + R_n, \qquad l \le n, \ k \le n - l,$$

$$-2E^{Q^n} [\ln M^n_{t_{k+l}} - \ln M^n_{t_k} | \mathcal{F}^n_{t_k}] \ge (l/n) T \sigma(\sigma - 2\lambda) - R_n, \qquad l \le n, \ k \le n - l,$$

where the positive sequence $R_n = O(n^{-1/2})$ does not depend on k and l.

Proof. The proof of the lemma stands on the following two estimations:

$$(4.10) \qquad \sup_{k \le n} \left| E^{Q^n} [2(\Delta \ln M^n_{t_k}) + (\Delta \ln M^n_{t_k})^2 | \mathcal{F}^n_{t_{k-1}}] \right| = O(n^{-3/2}),$$

$$(4.11) \qquad \sup_{k \le n} \left| E^{Q^n} [(\Delta \ln M^n_{t_k} + Y^n_{t_{k-1}})^2 - (Y^n_{t_k})^2 | \mathcal{F}^n_{t_{k-1}}] \right.$$

$$\left. - \sigma^n(\sigma^n + 2E^{Q^n} [Y^n_{t_k} \xi_k | \mathcal{F}^n_{t_{k-1}}]) \right| = O(n^{-3/2}),$$

where

$$Y^n := \ln M^n - \ln S^{n2} - \frac{\lambda^{n12} - \lambda^{n21}}{2} .$$

We start proving (4.10). Define the function

$$\Phi_2(z) := \ln(1 + z) - z + z^2/2 = O(z^3), \qquad z \to 0.$$

We get the following obvious identity:

$$2\Delta \ln M_{t_k}^n - 2\Delta L_{t_k}^n + (\ln M_{t_k}^n - E^{Q^n}[\ln M_{t_k}^n|\mathcal{F}_{t_{k-1}}^n])^2$$
$$= 2\Phi_2(\Delta L_{t_k}^n) + (\ln M_{t_k}^n - E^{Q^n}[\ln M_{t_k}^n|\mathcal{F}_{t_{k-1}}^n] - \Delta L_{t_k}^n)^2$$
$$+ 2\Delta L_{t_k}^n(\ln M_{t_k}^n - E^{Q^n}[\ln M_{t_k}^n|\mathcal{F}_{t_{k-1}}^n] - \Delta L_{t_k}^n).$$

Due to Lemma 4.1, we have the following asymptotics

$$\sup_{k \le n} \left|\ln M_{t_k}^n - E^{Q^n}[\ln M_{t_k}^n|\mathcal{F}_{t_{k-1}}^n] - \Delta L_{t_k}^n\right| = O(n^{-1}),$$

$$\|\Phi_2(\Delta L^n)\|_T = O(n^{-3/2}).$$

Using this, we get

$$\sup_{k \le n} \left|E^{Q^n}[2(\Delta \ln M_{t_k}^n) + (\ln M_{t_k}^n - E^{Q^n}[\ln M_{t_k}^n|\mathcal{F}_{t_{k-1}}^n])^2|\mathcal{F}_{t_{k-1}}^n]\right| = O(n^{-3/2}).$$

This relation in conjunction with (4.3) and (4.6), gives us the first asymptotic bound (4.10).

We recall the following bounds

$$-\lambda^{n21} \le -\ln(1 + \lambda^{n21}) \le \ln M^n - \ln S^{n2} \le \ln(1 + \lambda^{n12}) \le \lambda^{n12}.$$

Using this, we obtain that

(4.12) $$\|Y^n\|_T \le \lambda^n,$$

where $\lambda^n = \sqrt{T/n}\,\lambda$. By the relation

$$Y_{t_{k-1}}^n + \Delta \ln M_{t_k}^n = Y_{t_k}^n + \ln(1 + \mu^n + \sigma^n \xi_k),$$

we get the second main relation (4.11).

Now, we use (4.10) and (4.11) to complete the proof. With the expression

$$2\Delta \ln M_{t_k}^n + \Delta(Y_{t_k}^n)^2$$
$$= [2\Delta \ln M_{t_k}^n + (\Delta \ln M_{t_k}^n)^2] - [(\Delta \ln M_{t_k}^n + Y_{t_{k-1}}^n)^2 - (Y_{t_k}^n)^2]$$
$$+ 2Y_{t_{k-1}}^n \Delta \ln M_{t_k}^n,$$

we deduce from (4.6), (4.10), (4.11), and (4.12) the key relation

$$\sup_{k \le n} \left|E^{Q^n}[2\Delta \ln M_{t_k}^n + \Delta(Y_{t_k}^n)^2|\mathcal{F}_{t_{k-1}}^n]\right.$$
$$\left. + \sigma^n(\sigma^n + 2E^{Q^n}[Y_{t_k}^n \xi_k|\mathcal{F}_{t_{k-1}}^n])\right| = O(n^{-3/2}).$$

It remains to observe that

$$|2\sigma^n Y_{t_k}^n \xi_k| \le 2\sigma^n \lambda^n, \qquad k \le n.$$

Hence there exists a positive constant κ such that

$$-l\sigma^n(\sigma^n + 2\lambda^n) - \kappa ln^{-3/2}$$
$$\le 2E^{Q^n}[\ln M_{t_{k+l}}^n - \ln M_{t_k}^n|\mathcal{F}_{t_k}^n] + E^{Q^n}[(Y_{t_{k+l}}^n)^2|\mathcal{F}_{t_k}^n] - (Y_{t_k}^n)^2$$
$$\le -l\sigma^n(\sigma^n - 2\lambda^n) + \kappa ln^{-3/2}.$$

Using (4.12) and the inequality $ln^{-3/2} \le n^{-1/2}$, we get

$$-l\sigma^n(\sigma^n + 2\lambda^n) - \kappa n^{-1/2} - (\lambda^n)^2$$
$$\le 2E^{Q^n}[\ln M_{t_{k+l}}^n - \ln M_{t_k}^n|\mathcal{F}_{t_k}^n]$$
$$\le -l\sigma^n(\sigma^n - 2\lambda^n) + \kappa n^{-1/2} + (\lambda^n)^2.$$

This completes the proof. □

Lemma 4.4. *Let* $Z^n \in M^n$ *and let* $Q^n := Z_T^{n1}P$. *For each cluster point* Q *of the sequence* $\mathcal{L}(Z^n/Z^{n1}|Q^n)$, *there exists a process* $Z \in M$ *with* $Q = \mathcal{L}(Z)$.

Proof. Setting $\tilde{Q}^n = \mathcal{L}((1, M^n)|Q^n)$, Lemma 4.2 asserts that each cluster point Q of the tight sequence $\tilde{Q}^n$ charges only $\{1\} \times \mathbb{C}(\mathbb{R})$. On this set, the canonical process $\{(1, w_t); t \in [0, T]\}$ is a martingale under Q with respect to its natural filtration because of (4.8), see [5]. We shall show that the quadratic characteristics of logarithm of its second component is absolute continuous (with respect to Lebesgue measure) Q-a.s., with the bounds

$$(4.13) \qquad \sigma(\sigma - 2\lambda)dt \le d\langle \ln w\rangle_t \le \sigma(\sigma + 2\lambda)dt.$$

Equivalently, since $\{w_t; t \in [0, T]\}$ is a Q-martingale, $\langle \ln w\rangle$ is the bounded variation part of the semi-martingale $\{-2\ln w_t; t \in [0, T]\}$ and we show that

$$\sigma(\sigma - 2\lambda)E^Q \int_0^T g_t(w)dt$$
$$\le E^Q \int_0^T g_t(w)d\langle \ln w\rangle_t$$
$$\le \sigma(\sigma + 2\lambda)E^Q \int_0^T g_t(w)dt,$$

for any function $g : [0, T] \times \mathbb{D}(\mathbb{R}) \to \mathbb{R}_+$ which is bounded, continuous in the product of the usual topology on $[0, T]$ and the Skorohod topology on $\mathbb{D}(\mathbb{R})$ and adapted, i.e. $g_t(w)$ is $\sigma\{w_s, s \le t\}$-measurable for any t. The claim follows from Lemma 4.3 and (4.8). We have :

$$\limsup_{n \to \infty} E^{\tilde{Q}^n} g_s(w)(-2(\ln w_t - \ln w_s) - \sigma(\sigma + 2\lambda)(t - s)) \le 0,$$

and

$$\liminf_{n \to \infty} E^{\tilde{Q}^n} g_s(w)(-2(\ln w_t - \ln w_s) - \sigma(\sigma - 2\lambda)(t - s)) \ge 0.$$

Which lead to

$$-2E^Q g_s(w)(\ln w_t - \ln w_s) \le E^Q g_s(w)\sigma(\sigma + 2\lambda)(t - s),$$

and

$$-2E^Q g_s(w)(\ln w_t - \ln w_s) \ge E^Q g_s(w)\sigma(\sigma - 2\lambda)(t - s).$$

Hence Q on $\mathbb{C}(\mathbb{R}^2)$ is such that the (continuous) martingale part of $\{\ln w_t; \ t \in [0, T]\}$ has a quadratic characteristic process $\langle \ln w \rangle$ satisfying (4.13). From [7], Theorem 3.4.2, Q admits the following standard representation. There exist B, a standard Brownian motion under a probability v, and an adapted process g such that

$$\sigma(\sigma - 2\lambda) \le g^2 \le \sigma(\sigma + 2\lambda), \qquad 1 \le i \le d,$$

and

$$\mathscr{L}(1, \mathcal{E}(g \cdot B)| v) = Q.$$

$\square$

Construction of discrete martingales

The aim of the following section is to show that processes of $\mathcal{M}$ can be approximated by consistent price systems in $\mathcal{M}^n$. The following lemma gives a constructive way of approximating the martingales of a subset of $\mathcal{M}$.

Lemma 4.5. *Let B be a Brownian motion. Let g be an adapted continuous bounded function : $[0, T] \times \mathbb{D}(\mathbb{R}) \to \mathbb{R}^+ \setminus \{0\}$ such that, for some $\delta > 0$,*

$$(4.14) \qquad \delta \vee \sigma(\sigma - 2\lambda) + \delta \le g^2 \le \sigma(\sigma + 2\lambda) - \delta,$$
$$(4.15) \qquad |g_t(w) - g_s(v)| \le \kappa(|t - s| + \|w - v\|_T),$$

for $t, s \in [0, T]$, $v, w \in \mathbb{C}(\mathbb{R})$. Define the martingale

$$M = \mathcal{E}(g(B) \cdot B).$$

Then there exists a sequence $Z^n \in \mathcal{M}^n$ such that

$$\mathscr{L}(Z^n/Z^{1n}|Q^n) \to \mathscr{L}((1, M)|Q),$$

with $Q^n = Z_T^{1n} P$.

Proof. We consider the piecewise constant process

$$M_{t_k}^n = \frac{1 + 1/2\lambda^{n12}}{1 + 1/2\lambda^{n21}}\left(1 + K_{t_k}^n \sqrt{T/n}\xi_k\right)S_{t_k}^{n2}, \qquad 0 \le k \le n,$$

with K^n the predictable process defined by

(4.16)
$$K_{t_k}^n = \frac{1}{2\sigma}(g_{t_k}^n)^2 - \frac{\sigma}{2},$$

$$g_{t_k}^n = g_{t_{k-1}}((B_{t_l}^n)_{l=0}^{k-1}),$$

where the process B^n is piecewise constant with the jumps

(4.17)
$$\Delta B_{t_k}^n = (g_{t_k}^n)^{-1}\Delta L_{t_k}^n,$$

$$\Delta L_{t_k}^n = (M_{t_{k-1}}^n)^{-1}\Delta M_{t_k}^n.$$

The proof consists in two steps. The first one is to construct from M^n a sequence of consistent price systems in $\mathcal{M}^n$. The second step is to check the convergence.

According to (4.14),

$$-\lambda + \varepsilon \le K^n \le \lambda - \varepsilon,$$

for some $\varepsilon > 0$. Using the Taylor expansion formulae, we get the bounds

$$1 - \lambda^{n21} + \varepsilon R_n^1 \le \frac{1 + 1/2\lambda^{n12}}{1 + 1/2\lambda^{n21}}\left(1 + K_{t_k}^n \sqrt{T/n}\xi_k\right) \le 1 + \lambda^{n12} - \varepsilon R_n^1,$$

where $R_n^1 = O(n^{-1/2})$ and $R_n^1 > 0$ for large n. It is easily seen that

$$\frac{1}{1 + \lambda^{n21}}S^{n2} \le M^n \le (1 + \lambda^{n12})S^{n2}$$

for sufficiently large n. These inequalities show that $(1, M^n)$ takes values in $\widehat{K}^{\Lambda^{n*}} \setminus \{0\}$ for sufficiently large n. Our aim now is to determine the martingale measure of M^n. We compute the stochastic logarithm of M^n,

$$\Delta L_{t_k}^n = \frac{M_{t_k}^n}{M_{t_{k-1}}^n} - 1$$

$$= \frac{(1 + \mu^n + \sigma^n\xi_k)(1 + K_{t_k}^n \sqrt{T/n}\xi_k)}{1 + K_{t_{k-1}}^n \sqrt{T/n}\xi_{k-1}} - 1$$

$$= \sqrt{\frac{T}{n}}\frac{(\sigma + K_{t_k}^n + \mu_n K_{t_k}^n)\xi_k + \mu\sqrt{T/n} + \sigma_n K_{t_k}^n - K_{t_{k-1}}^n \xi_{k-1}}{1 + \sqrt{T/n}K_{t_{k-1}}^n \xi_{k-1}}.$$

Observe that M^n is a Q^n-martingale where Q^n is given by

$$Q^n = \mathcal{E}(q^n)_T P, \qquad \Delta q_{t_k}^n = -\frac{\mu\sqrt{T/n} + \sigma_n K_{t_k}^n - K_{t_{k-1}}^n \xi_{k-1}}{(\sigma + K_{t_k}^n + \mu_n K_{t_k}^n)}\xi_k,$$

recalling that for a piecewise constant process q,

$$\mathcal{E}(q)_t = \prod_{s \leq t}(1 + \Delta q_s).$$

Setting

$$Z_t^n = E\left[\mathcal{E}(q^n)_T | \mathcal{F}_{t_k}^n\right](1, M_{t_k}^n), \qquad t_k \leq t < t_{k+1},$$

we get a sequence of martingales taking values in $\widehat{K}^{\Lambda^n*}$, that is a sequence of consistent price systems.

In view of (4.17), we have the expression

$$M^n = \mathcal{E}(g^n \cdot B^n).$$

We shall use a version of the Central Limit Theorem to show the convergence of $\mathscr{L}(B^n | Q^n)$ to the law of a Brownian motion. We need to compute the increments of the quadratic variation process of B^n, that is $E^{Q^n}[(\Delta B_{t_k}^n)^2 | \mathcal{F}_{t_{k-1}}^n]$. First, according to (4.15) and (4.17), observe that

$$\|\Delta B^n\|_T = O(n^{-1/2}), \qquad \|\Delta K^n\|_T = O(n^{-1/2}).$$

It follows that

$$(4.18) \qquad \sup_{k \leq n}\left|\Delta L_{t_k}^n - \sqrt{T/n}[(\sigma + K_{t_k}^n)\xi_k - K_{t_k}^n\xi_{k-1}]\right| = O\left(n^{-1}\right),$$

and

$$(4.19) \qquad \sup_{k \leq n}\left|\Delta q_{t_k}^n - \frac{K_{t_k}^n\xi_{k-1}\xi_k}{\sigma + K_{t_k}^n}\right| = O\left(n^{-1/2}\right).$$

Having in mind the expression

$$E^{Q^n}[(\Delta B_{t_k}^n)^2 | \mathcal{F}_{t_{k-1}}^n] = (g_k^n)^{-2}E[(1 + \Delta q_{t_k}^n)(\Delta L_{t_k}^n)^2 | \mathcal{F}_{t_{k-1}}^n],$$

we deduce from (4.18) and (4.19),

$$E^{Q^n}[(\Delta B_{t_k}^n)^2 | \mathcal{F}_{t_{k-1}}^n] = \frac{T}{n}(g_k^n)^{-2}\left((\sigma + K_{t_k}^n)^2 - (K_{t_k}^n)^2\right) + R_{t_k}^n,$$

where $\|R^n\|_T = O(n^{-3/2})$. Finally, with the definition of K^n, (4.16), it is easily seen that

$$E^{Q^n}[(\Delta B_{t_k}^n)^2 | \mathcal{F}_{t_{k-1}}^n] = \frac{T}{n} + R_{t_k}^n.$$

Note also that the sequence B^n satisfies the conditional Lindeberg hypothesis, Property VIII.3.31 in [5]. By the use of the Central Limit Theorem, [5], VIII.3.33, we get the existence of a Brownian motion B such that

$$\mathscr{L}(B^n, g(B^n) | Q^n) \to \mathscr{L}(B, g(B)).$$

The announced convergence can be checked through the convergence of the stochastic exponential, and then the convergence of $\mathscr{L}(M^n|Q^n)$ to the law of the process M holds. □

Note that approximating processes of Lemma 4.5 allows us to approximate processes of $\mathcal{M}$. Indeed, let $Z \in \mathcal{M}$, $Z^2 = \mathcal{E}(g \cdot B)$. It is easily seen that we can construct a sequence of functions $(g^m)_{m \in \mathbb{N}}$ satisfying the assumptions of Lemma 4.5 with

$$E \int \left| g_t - g_t^m(B) \right|^2 dt \to 0.$$

Using Burkholder's inequality, we get that

$$E \left\| \mathcal{E}(g \cdot B) - \mathcal{E}(g_-^m(B) \cdot B) \right\|_T \to 0.$$

5. Proof of the main results

Preliminary remarks

We first give some general remarks and tools which link the technical ideas from Section 4 with super hedging issues.

Remind the assertion (4.2), that is for any $Z \in \mathcal{M}^n$,

(5.20)
$$\frac{1 - \kappa^n}{1 + \kappa^n} \leq Z_0^2 \leq \frac{1 + \kappa^n}{1 - \kappa^n},$$

and, more generally,

$$\frac{1 - \kappa^n}{1 + \kappa^n} S^{n2} \leq Z^2/Z^1 \leq \frac{1 + \kappa^n}{1 - \kappa^n} S^{n2}.$$

Now we show that the particular convergence described in Lemmas 4.2 and 4.5 is consistent with the hedging theorem. Let $Z^n \in \mathcal{M}^n$ be such that for $M^n := Z^{n2}/Z^{n1}$ and $Q^n := Z_T^{n1} P$ we have

$$\mathscr{L}\left((1, M^n)| Q^n\right) \to \mathscr{L}(Z),$$

for some $Z \in \mathcal{M}$. It follows from Lemma 4.2.2 that for any $v \in \mathbb{R}^2$,

(5.21)
$$EZ_T^n(F(S^n) - v) \to EZ_T(F(Z) - v),$$

since

$$EZ_T^n(F(S^n) - v) = E^{Q^n}(1, M_T^n)(F(S^n) - v).$$

We end this paragraph observing the fact that increasing the initial capital both on the first and the second asset helps to hedge the European option. Indeed, for each $v, \delta > 0$, $Z^n \in \mathcal{M}^n$, we have

(5.22)
$$EZ_T^n(F(S^n) - (v + \delta\mathbf{1})) \leq EZ_T^n(F(S^n) - v) - 2\frac{1 - \kappa^n}{1 + \kappa^n}\delta.$$

Moreover, this bound is uniform on the choice of the consistent price system.

Proof of Theorem 3.2

The proof of this theorem is similar to the one given in [8]. Note that

$$x_n = \sup_{Z \in \mathcal{M}^n} EZ_T F(S^n),$$

and

$$x = \sup_{Z \in \mathcal{M}} EZ_T F(Z).$$

We proceed by establishing the following two inequalities:

$$\limsup_n x^n \leq x, \qquad \liminf_n x^n \geq x.$$

For the first one, we fix the sequence $Z^n \in \mathcal{M}^n$ such that

$$EZ_T^n F(S^n) \geq x^n - 1/n.$$

According to Lemmas 4.2 and 4.4, there exist a subsequence Z^{n_k} and a process $Z \in \mathcal{M}$ such that

$$\limsup_n EZ_T^n F(S^n) = \lim_k EZ_T^{n_k} F(S^{n_k}) = EZ_T F(Z) \leq x.$$

Conversely, we fix $\varepsilon > 0$ and choose $Z \in \mathcal{M}$ such that

$$EZ_T F(Z) \geq x - \varepsilon.$$

By virtue of Lemma 4.5, there exists a sequence $Z^n \in \mathcal{M}^n$ such that

$$\liminf_n EZ_T^n F(S^n) = EZ_T F(Z).$$

Since ε is arbitrary, we get $\liminf x^n \geq x$, and Theorem 3.2 is proved.

Proof of Theorem 3.1, Assertion 1

The proof of Theorem 3.1 follows the same reasoning based on choosing the best candidate between the consistent price systems. However, the fact that we consider convergence of sets makes the demonstration more involved. Here we prove the first assertion.

Fix $v \in \Gamma$, we shall construct a sequence $v^n \in \Gamma^n$ such that $v^n \to v$. Choose a sequence $Z^n \in \mathcal{M}^n$ such that

$$EZ_T^n(F(S^n) - v) + \frac{1}{n} \geq \sup_{Z \in \mathcal{M}^n} EZ_T(F(S^n) - v).$$

As a consequence of Lemmas 4.2 and 4.4, there exists $Z \in \mathcal{M}$ such that

$$\limsup_n EZ_T^n(F(S^n) - v) = EZ_T(F(Z) - v) \leq 0.$$

It follows that there is a positive sequence $\delta^n \to 0$ such that

$$EZ_T^n(F(S^n) - v) \leq \delta^n.$$

Define v^n by increasing the initial capital v to

$$v^n = v + \frac{1}{2} \frac{1 + \kappa^n}{1 - \kappa^n} \left(\delta^n + \frac{1}{n} \right) \mathbf{1}.$$

Having in mind (5.22), it is easily seen that for any $Z \in \mathcal{M}^n$, we have:

$$EZ_T(F(S^n) - v^n) \leq EZ_T(F(S^n) - v) - \left(\delta^n + \frac{1}{n} \right) \leq EZ_T^n(F(S^n) - v) - \delta^n \leq 0.$$

So we constructed the desired sequence $v^n \in \Gamma^n$ such that $v^n \to v$.

Proof of Theorem 3.1, Assertion 2

It remains to show that for a convergent (sub)sequence $v^n \in \Gamma^n$, the limit v belongs to Γ. Fix $\varepsilon > 0$ and choose $Z \in \mathcal{M}$ such that

$$EZ_T(F(Z) - v) \geq \sup_{Z \in \mathcal{M}} EZ_T(F(Z) - v) - \varepsilon.$$

By virtue of Lemma 4.5 and (5.21), there is a sequence $Z^n \in \mathcal{M}^n$ such that

$$\liminf_n EZ_T^n(F(S^n) - v) = EZ_T(F(Z) - v).$$

Note that

$$\liminf_n EZ_T^n(F(S^n) - v) = \liminf_n EZ_T^n(F(S^n) - v^n) + \liminf_n Z_0^n(v^n - v),$$

and

$$\liminf_n Z_0^n(v^n - v) = 0,$$

since Z_0^n is bounded, (5.20). We can conclude that

$$EZ_T(F(Z) - v) \leq 0$$

and since ε is arbitrary, v belongs to Γ. This ends the proof. $\square$

Acknowledgement

I gratefully thank the organizers of the International Workshop on Finance 2012 for their warm hospitality and their generosity. I also express my thanks to Professor Yuri Kabanov for his very careful reading of the paper and valuable remarks. Financial support by Conseil Régional de Franche-Comté is appreciated.

References

1. Darses S. and Lépinette E. Limit Theorem for a Modified Leland Hedging Strategy under Constant Transaction Costs rate. "The Musiela Festschrift ". Springer.
2. Grépat J. On a Multi-Asset Version of the Kusuoka Limit Theorem of Option Replication under Transaction Costs. Submitted.
3. Grépat J. and Kabanov Yu. Small transaction costs, absence of arbitrage and consistent price systems. Finance and Stochastics, 16 (2012), 3, 357-368.
4. Hildenbrand, W. *Core and equilibria of a large economy*. Princeton University Press, Princeton, N.J.,1974.
5. Jacod, J. and Shiryaev, A. N. *Limit theorems for stochastic processes*, Springer, Berlin, 2nd edition, 2003.
6. Kabanov, Y. and Safarian, M. *Markets with transaction costs*, Springer Finance, Springer, Berlin, 2009.
7. Karatzas, I. and Shreve, S. E. *Brownian motion and stochastic calculus*, Graduate Texts in Mathematics, Springer, Berlin, second edition, 1991.
8. Kusuoka, S. Limit theorem on option replication cost with transaction costs, Ann. Appl. Probab., 5 (1995), 1, 198–221.
9. Leland, H. E. Option pricing and replication with transaction costs, J. Finance, 40 (1985), 5, 1283–1301.
10. Lépinette E. Modified Leland's Strategy for Constant Transaction Costs Rate. Mathematical Finance.

Optimal Execution for Uncertain Market Impact: Derivation and Characterization of a Continuous-Time Value Function

Kensuke Ishitani[1] and Takashi Kato[2]

[1]Department of Mathematics, Meijo University, Tempaku, Nagoya 468-8502, Japan
[2]Division of Mathematical Science for Social Systems, Graduate School of Engineering Science, Osaka University, 1-3 Machikaneyama-cho, Toyonaka, Osaka 560-8531, Japan
Email:kishitani@meijo-u.ac.jp   kato@sigmath.es.osaka-u.ac.jp

In this paper, we study an optimal execution problem in the case of un-certainty in market impact to derive a more realistic market model. Our model is a generalized version of that in [6], where a model of optimal ex-ecution with deterministic market impact was formulated. First, we con-struct a discrete-time model as a value function of an optimal execution problem. We express the market impact function as a product of a deter-ministic part (an increasing function with respect to the trader's execution volume) and a noise part (a positive random variable). Then, we derive a continuous-time model as a limit of a discrete-time value function. We find that the continuous-time value function is characterized by an optimal control problem with a Lévy process and investigate some of its proper-ties, which are mathematical generalizations of the results in [6]. We also consider a typical example of the execution problem for a risk-neutral trader under log-linear/quadratic market impact with Gamma-distributed noise.

Key words: market liquidity, optimal execution, uncertain market impact, Lévy process, viscosity solution, Hamilton–Jacobi–Bellman equation

1. Introduction

The optimal portfolio management problem is central in mathematical finance theory. There are various studies on this problem, and recently more realistic problems, such as liquidity problems, have attracted considerable attention. In this paper, we focus on market impact (MI), which is the effect of the investment behavior of traders on security prices. MI plays an important role in portfolio theory, and is also significant when we consider the case of an optimal execution problem, where a trader has a certain amount of security holdings (shares of a security held) and attempts to liquidate them before the time horizon. The optimal execution problem with MI has been studied in several papers ([1], [2], [3], [4], [10] and references therein,) and in [6] such a problem is formulated mathematically.

It is often assumed that the MI function is deterministic. This assumption means that we can obtain information about MI in advance. However, in a real market it is difficult to estimate the effects of MI. Moreover, it often happens that a high concentration of unexpected orders will result in overfluctuation of the price. The Flash Crash in the United States stock market is a notable precedent of unusual thinning liquidity: On May 6th, 2010, the Dow Jones Industrial Average plunged by about 9%, only to recover the losses within minutes. Considering the uncertainty in MI, it is thus more realistic and meaningful to construct a mathematical model of random MI. Moazeni et al. [9] studied the uncertainty in MI caused by other institutions by compound Poisson processes, and then studied an optimization problem of expected proceeds of execution in a discrete-time setting. They considered the uncertainty in arrival times of large trades from other institutions; however, MI functions of decision makers themselves were given as deterministic linear functions so that the decision makers knew how their own execution affected the market price of the security (the coefficients of MI functions were regarded as "expected price depressions caused by trading assets at a unit rate").

In this paper, we generalize the framework in [6], particularly considering a random MI function. We follow the approach in [6]: we construct a discrete-time model and take a limit to derive a continuous-time model and a corresponding value function. We assume that noise in the MI function in a discrete-time model is independent of both time and trading volume. The randomness of MI in the continuous-time model is described as a jump of a Lévy process. We study some properties of the continuous-time value function which are mathematical generalizations of the results in [6]. In particular, we find that the value function is characterized as a viscosity solution of the corresponding Hamilton-Jacobi-Bellman equation (abbreviated as HJB) when the MI function is sufficiently strong. We also perform a comparison with the case of deterministic MI and show that noise in MI makes a risk-neutral trader underestimate the MI cost. This means that a trader attempting to minimize the expected liquidation cost is not particularly sensitive to the uncertainty in MI. Moreover, we present generalizations of examples in [6] and investigate the effects of noise in MI on the optimal strategy of a trader by numerical experiments. We consider a risk-neutral trader execution problem with a log-linear/quadratic MI function whose noise is given by the Gamma distribution.

The rest of this paper is organized as follows. In Section 2, we present the mathematical formulation of our model. We set a discrete-time model of an optimal execution problem as our basic model and define the corresponding value function. In Section 3, we present our results, showing that the continuous-time value function is derived as a limit of the discrete-time one. Moreover, we investigate the continuity of the derived value function. Note that the results in this section are of the same form as those in [6]. In Section 4, we study the characterization of the value function as a viscosity solution of the corresponding HJB equation as a direct consequence of the result in [6]. In Section 5, we consider the case where the trader must sell all the shares of the security, which is referred to as a "sell-off condition." We also study the optimization problem under the sell-off condition and show that the results in Section 4 in [6] also hold in our model. Section 6 treats the comparison between deterministic MI and random MI in a risk-neutral framework. In Section 7 we present some examples based on the proposed model. We conclude this paper in Section 8.

2. The Model

In this section, we present the details of the proposed model. Let $(\Omega, \mathcal{F}, P)$ be a complete probability space. $T > 0$ denotes a time horizon, and we assume $T = 1$ for brevity. We assume that the market consists of one risk-free asset (cash) and one risky asset (a security). The price of cash is always 1, which means that a risk-free rate is zero. The price of the security fluctuates according to a certain stochastic flow, and is influenced by sales performed by traders.

First, we consider a discrete-time model with a time interval $1/n$. We consider a single trader who has an endowment of $\Phi_0 > 0$ shares of a security. This trader liquidates the shares Φ_0 over a time interval $[0, 1]$ considering the effects of MI with noise. We assume that the trader sells shares at only times $0, 1/n, \ldots, (n-1)/n$ for $n \in \mathbb{N} = \{1, 2, 3, \ldots\}$.

For $l = 0, \ldots, n$, we denote by S_l^n the price of the security at time l/n, and we also denote $X_l^n = \log S_l^n$. Let $s_0 > 0$ be an initial price (i.e., $S_0^n = s_0$) and $X_0^n = \log s_0$. If the trader sells an amount ψ_l^n at time l/n, the log price changes to $X_l^n - g_l^n(\psi_l^n)$, and by this execution (selling) the trader obtains an amount of cash $\psi_l^n S_l^n \exp(-g_l^n(\psi_l^n))$ as proceeds. Here, the random function

$$g_l^n(\psi, \omega) = c_l^n(\omega) g_n(\psi), \quad \psi \in [0, \Phi_0], \; \omega \in \Omega$$

denotes MI with noise, which is given by the product of a positive random variable c_l^n and a deterministic function $g_n : [0, \Phi_0] \longrightarrow [0, \infty)$. The function g_n is assumed to be non-decreasing, continuously differentiable and satisfying $g_n(0) = 0$. Moreover, we assume that $(c_l^n)_l$ is i.i.d., and therefore noise in MI is time-homogeneous. Note that if c_l^n is a constant (i.e., $c_l^n \equiv c$ for some $c > 0$,) then this setting is the same as in [6].

After trading at time l/n, X_{l+1}^n and S_{l+1}^n are given by

$$(1) \qquad X_{l+1}^n = Y\left(\frac{l+1}{n}; \frac{l}{n}, X_l^n - g_l^n(\psi_l^n)\right), \; S_{l+1}^n = e^{X_{l+1}^n},$$

where $Y(t; r, x)$ is the solution of the following stochastic differential equation (SDE) on the filtered space $(\Omega, \mathcal{F}, (\mathcal{F}_t^B)_t, P)$

$$\begin{cases} dY(t; r, x) = \sigma(Y(t; r, x))dB_t + b(Y(t; r, x))dt, t \geq r, \\ Y(r; r, x) = x, \end{cases}$$

where $(B_t)_{0 \leq t \leq 1}$ is standard one-dimensional Brownian motion which is independent of $(c_l^n)_l$, $(\mathcal{F}_t^B)_t$ is its Brownian filtration, and $b, \sigma : \mathbb{R} \longrightarrow \mathbb{R}$ are Borel functions. We assume that b and σ are bounded and Lipschitz continuous. Then, for each $r \geq 0$ and $x \in \mathbb{R}$, there exists a unique solution.

At the end of the time interval $[0, 1]$, the trader has an amount of cash W_n^n and an amount of the security φ_n^n, where

(2) $$W_{l+1}^n = W_l^n + \psi_l^n S_l^n e^{-g_l^n(\psi_l^n)}, \quad \varphi_{l+1}^n = \varphi_l^n - \psi_l^n$$

for $l = 0, \ldots, n - 1$ and $W_0^n = 0$, $\varphi_0^n = \Phi_0$. We say that an execution strategy $(\psi_l^n)_{l=0}^{n-1}$ is admissible if $(\psi_l^n)_l \in \mathcal{A}_n^n(\Phi_0)$ holds, where $\mathcal{A}_k^n(\varphi)$ is the set of strategies $(\psi_l^n)_{l=0}^{k-1}$ such that ψ_l^n is $\mathcal{F}_l^n = \sigma\{(B_t)_{t \leq l/n}, c_0^n, \cdots, c_{l-1}^n\}$-measurable, $\psi_l^n \geq 0$ for each $l = 0, \ldots, k - 1$ and $\sum_{l=0}^{k-1} \psi_l^n \leq \varphi$ almost surely.

Then, the investor's problem is to choose an admissible strategy to maximize the expected utility $E[u(W_n^n, \varphi_n^n, S_n^n)]$, where $u \in C$ is the utility function employed by the investor and C is the set of non-decreasing, non-negative and continuous functions on $D = \mathbb{R} \times [0, \Phi_0] \times [0, \infty)$ such that

(3) $$u(w, \varphi, s) \leq C_u(1 + |w|^{m_u} + s^{m_u}), \quad (w, \varphi, s) \in D$$

for some constants $C_u, m_u > 0$.

For $k = 1, \ldots, n$, $(w, \varphi, s) \in D$ and $u \in C$, we define the (discrete-time) value function $V_k^n(w, \varphi, s; u)$ by

$$V_k^n(w, \varphi, s; u) = \sup_{(\psi_l^n)_{l=0}^{k-1} \in \mathcal{A}_k^n(\varphi)} E[u(W_k^n, \varphi_k^n, S_k^n)]$$

subject to (1) and (2) for $l = 0, \ldots, k - 1$ and $(W_0^n, \varphi_0^n, S_0^n) = (w, \varphi, s)$ (for $s = 0$, we set $S_l^n \equiv 0$). We denote such a triplet of processes $(W_l^n, \varphi_l^n, S_l^n)_{l=0}^k$ by $\Xi_k^n(w, \varphi, s; (\psi_l^n)_l)$, and denote $V_0^n(w, \varphi, s; u) = u(w, \varphi, s)$. Then, this problem is equivalent to $V_n^n(0, \Phi_0, s_0; u)$. We consider the limit of the value function $V_k^n(w, \varphi, s; u)$ as $n \to \infty$.

Let $h : [0, \infty) \longrightarrow [0, \infty)$ be a non-decreasing continuous function. We introduce the following condition for $g_n(\psi)$.

[A] $\lim_{n \to \infty} \sup_{\psi \in [0, \Phi_0]} \left| \frac{d}{d\psi} g_n(\psi) - h(n\psi) \right| = 0$.

Moreover, we assume the following conditions for $(c_l^n)_l$.

[B1] Define $\gamma_n = \text{essinf}_\omega \, c_l^n(\omega)$. For any $n \in \mathbb{N}$, it holds that $\gamma_n > 0$. In addition,

(4)
$$\frac{h(x/\gamma_n)}{n} \longrightarrow 0, \quad n \to \infty$$

holds for $x \geq 0$.

[B2] Let μ_n be the distribution of $\frac{c_0^n + \ldots + c_{n-1}^n}{n}$. Then, μ_n has a weak limit μ as $n \to \infty$.

[B3] There is a sequence of infinitely divisible distributions $(p_n)_n$ on $\mathbb{R}$ such that $\mu_n = \mu * p_n$ and either

    [B3-a] $\int_{\mathbb{R}} x^2 p_n(dx) = O(1/n^3)$ as $n \to \infty$

    or

    [B3-b] There is a sequence $(K_n)_n \subset (0, \infty)$ such that $K_n = O(1/n)$, $p_n((-\infty, -K_n)) = 0$ (or $p_n((K_n, \infty)) = 0$) and $\int_{\mathbb{R}} x p_n(dx) = O(1/n)$ as $n \to \infty$, where O denotes the order notation (Landau's symbol).

Let us discuss condition [B1]. First, note that γ_n is independent of l, because $c_l^n, l = 0, 1, 2, \ldots$ are identically distributed. Next, we study when the convergence (4) holds. Since h is non-decreasing, we see that

$$\frac{h(x/\gamma_n)}{n} \leq \frac{h(\infty)}{n}, \quad n \in \mathbb{N},$$

where $h(\infty) = \lim_{\zeta \to \infty} h(\zeta) \in [0, \infty]$ (which is well-defined by virtue of the monotonicity of h). This inequality tells us that (4) is always fulfilled whenever $h(\infty) < \infty$. In the case of $h(\infty) = \infty$, we have the following example:

(5)
$$h(\zeta) = \alpha \zeta^p, \quad \gamma_n = \frac{1}{n^{1/p - \delta}} \quad (p, \delta > 0, \ \delta \leq 1/p).$$

We can actually confirm (4) by observing

$$\frac{h(x/\gamma_n)}{n} = \frac{\alpha x^p}{n^{p\delta}} \longrightarrow 0, \quad n \to \infty.$$

Here, the condition [B2] holds only when $\limsup_n \gamma_n < \infty$. Indeed, under [B2], we easily see that the support of the distribution μ is included in the interval $[\limsup_n \gamma_n, \infty)$. Note that γ_n of (5) satifies $\limsup_n \gamma_n \leq 1$ because of the relation $\delta \leq 1/p$.

Since μ is an infinitely divisible distribution, there is some Lévy process (subordinator) $(L_t)_{0 \leq t \leq 1}$ on a certain probability space such that L_1 is distributed by μ. Without loss of generality, we may assume that $(L_t)_t$ and $(B_t)_t$ are defined on the same filtered space. Since $(c_l^n)_l$ is independent of $(B_t)_t$, we may also assume that $(L_t)_t$ is independent of $(B_t)_t$. Let ν be the Lévy measure of $(L_t)_t$. We assume the following moment condition for ν.

[C] $\int_{(0,\infty)}(z + z^2)\nu(dz) < \infty$.

Now, we define the function which gives the limit of the discrete-time value function. For $t \in [0, 1]$ and $\varphi \in [0, \Phi_0]$ we denote by $\mathcal{A}_t(\varphi)$ the set of $(\mathcal{F}_r)_{0 \le r \le t}$-adapted and càglàd processes (i.e., left-continuous and having a right limit at each point) $(\zeta_r)_{0 \le r \le t}$ such that $\zeta_r \ge 0$ for each $r \in [0, t]$, $\int_0^t \zeta_r dr \le \varphi$ almost surely and

$$(6) \qquad \sup_{(r,\omega) \in [0,t] \times \Omega} \zeta_r(\omega) < \infty,$$

where $\mathcal{F}_r = \sigma\{B_v, L_v ; v \le r\} \vee \{\text{Null Sets}\}$. Here, the supremum in (6) is taken over all values in $[0, t] \times \Omega$. Note that we may replace sup in (6) with esssup.

For $t \in [0, 1]$, $(w, \varphi, s) \in D$ and $u \in C$, we define $V_t(w, \varphi, s; u)$ by

$$(7) \qquad V_t(w, \varphi, s; u) = \sup_{(\zeta_r)_r \in \mathcal{A}_t(\varphi)} \mathrm{E}[u(W_t, \varphi_t, S_t)]$$

subject to

$$dW_r = \zeta_r S_r dr,$$
$$d\varphi_r = -\zeta_r dr,$$
$$dX_r = \sigma(X_r) dB_r + b(X_r) dr - g(\zeta_r) dL_r, \quad S_r = \exp(X_r)$$

and $(W_0, \varphi_0, S_0) = (w, \varphi, s)$, where $\hat{\sigma}(s) = s\sigma(\log s), \hat{b}(s) = s\{b(\log s) + \frac{1}{2}\sigma(\log s)^2\}$ for $s > 0$ ($\hat{\sigma}(0) = \hat{b}(0) = 0$) and $g(\zeta) = \int_0^\zeta h(\zeta')d\zeta'$. We denote such a triplet of processes $(W_r, \varphi_r, S_r)_{0 \le r \le t}$ by $\Xi_t(w, \varphi, s; (\zeta_r)_r)$. Note that $V_0(w, \varphi, s; u) = u(w, \varphi, s)$. We call $V_t(w, \varphi, s; u)$ a continuous-time value function. Also note that $V_t(w, \varphi, s; u) < \infty$ for each $t \in [0, 1]$ and $(w, \varphi, s) \in D$.

3. Properties of Value Functions

First, we give the convergence theorem for value functions.

Theorem 3.1. *For each* $(w, \varphi, s) \in D$, $t \in [0, 1]$ *and* $u \in C$ *it holds that*

$$\lim_{n \to \infty} V^n_{[nt]}(w, \varphi, s; u) = V_t(w, \varphi, s; u),$$

where $[nt]$ *is the greatest integer* $\le nt$.

According to this theorem, a discrete-time value function converges to $V_t(w, \varphi, s; u)$ by shortening the time intervals of execution. This implies that we can regard $V_t(w, \varphi, s; u)$ as the value function of the continuous-time model of an optimal execution problem with random MI. This result is almost the same as in [6], with the exception that the term of MI is given as an increment $g(\zeta_r)dL_r$. Let

$$L_t = \gamma t + \int_0^t \int_{(0,\infty)} zN(dr, dz)$$

be the Lévy decomposition of $(L_t)_t$, where $\gamma \geq 0$ and $N(\cdot, \cdot)$ is a Poisson random measure. Then, $g(\zeta_r)dL_r$ can be divided into two terms as follows:

$$g(\zeta_{r-})dL_r = \gamma g(\zeta_r)dr + g(\zeta_{r-}) \int_{(0,\infty)} zN(dr, dz).$$

The last term on the right-hand side indicates the effect of noise in MI. This means that noise in MI appears as a jump of a Lévy process. Using the above representation and Itô's formula, we see that when $s > 0$, the process $(S_r)_r$ satisfies

$$dS_r = \hat{\sigma}(S_r)dB_r + \hat{b}(S_r)dr - \left\{ \gamma g(\zeta_r)S_r dr + S_{r-} \int_{(0,\infty)} (1 - e^{-g(\zeta_{r-})z})N(dz, dr) \right\}.$$

Regarding the continuity of the continuous-time value function, we have the following theorem.

Theorem 3.2. *Let $u \in C$.*
(i) *If $h(\infty) = \infty$, then $V_t(w, \varphi, s; u)$ is continuous in $(t, w, \varphi, s) \in [0, 1] \times D$.*
(ii) *If $h(\infty) < \infty$, then $V_t(w, \varphi, s; u)$ is continuous in $(t, w, \varphi, s) \in (0, 1] \times D$ and $V_t(w, \varphi, s; u)$ converges to $Ju(w, \varphi, s)$ uniformly on any compact subset of D as $t \downarrow 0$, where $Ju(w, \varphi, s)$ is given as*

$$\begin{cases} \sup_{\psi \in [0,\varphi]} u\left(w + \frac{1-e^{-\gamma h(\infty)\psi}}{\gamma h(\infty)}s, \varphi - \psi, se^{-\gamma h(\infty)\psi}\right) & (\gamma h(\infty) > 0), \\ \sup_{\psi \in [0,\varphi]} u(w + \psi s, \varphi - \psi, s) & (\gamma h(\infty) = 0). \end{cases}$$

This is also quite similar to the result in [6], whereby continuities in w, φ and s of the value function are always guaranteed, but continuity in t at the origin depends on the state of the function h at infinity. When $h(\infty) = \infty$, MI for large sales is sufficiently strong ($g(\zeta)$ diverges rapidly with $\zeta \to \infty$) to make a trader avoid instant liquidation: an optimal policy is "no trading" in infinitesimal time, and thus V_t converges to u as $t \downarrow 0$. When $h(\infty) < \infty$, the value function is not always continuous at $t = 0$ and has the right limit $Ju(w, \varphi, s)$. In this case, MI for large sales is not particularly strong ($g(\zeta)$ still diverges, but the divergence speed is low) and there is room for liquidation within infinitesimal time. The function $Ju(w, \varphi, s)$ corresponds to the utility of the liquidation of the trader, who sells part of the shares of a security ψ by dividing it infinitely within an infinitely short time (sufficiently short that the fluctuation in the price of the security can be ignored) and obtains an amount $\varphi - \psi$, that is,

(8) $$\zeta_r^\delta = \frac{\psi}{\delta}1_{[0,\delta]}(r), \quad r \in [0, t] \; (\delta \downarrow 0).$$

We pay attention to the fact that the jump part $g(\zeta_{r-}) \int_{(0,\infty)} zN(dr, dz)$ does not change the result. Note that if $\gamma = 0$ and $h(\infty) < \infty$, then the effect of MI disappears in $Ju(w, \varphi, s)$. This situation may occur even if $E[c_k^n] \geq \varepsilon_0$ (or $E[L_1] \geq \varepsilon_0$) for some $\varepsilon_0 > 0$.

For the proofs of Theorems 3.1–3.2, please refer to [5].

4. Characterization of the Value Function as a Viscosity Solution of a Corresponding HJB

As presented in [6], we can characterize our value function as a viscosity solution of a corresponding HJB when $h(\infty) = \infty$. First, we present the Bellman principle (dynamic programming principle). Let us define $Q_t : C \longrightarrow C$ by $Q_t u(w, \varphi, s) = V_t(w, \varphi, s; u)$. Then, we can easily confirm that Q_t is well-defined as a nonlinear operator. The same proof as that for Theorem 3 in [6] gives the following proposition.

Prop 4.1. For each $r, t \in [0, 1]$ with $t + r \leq 1$, $(w, \varphi, s) \in D$ and $u \in C$, it holds that $Q_{t+r} u(w, \varphi, s) = Q_t Q_r u(w, \varphi, s)$.

Note that this proposition is also required to prove Theorem 3.2. At this stage, we introduce the HJB corresponding to our value function. We start by formal calculation to derive the HJB and subsequently show that we can apply Theorems 4–5 in [6] directly to our case, in spite of the fact that our value function is a generalization of that in [6].

We begin by fixing $t \in (0, 1]$ and letting $h > 0$ be such that $t - h \geq 0$. Then, Proposition 4.1 leads us to

$$V_t(w, \varphi, s; u) = Q_h Q_{t-h} u(w, \varphi, s; u) = \sup_{(\zeta_r)_r \in \mathcal{A}_h(\varphi)} E[V_{t-h}(W_h, \varphi_h, S_h; u)].$$

If we restrict the admissible strategies to constant ones and assume the smoothness of $V_t(w, \varphi, s; u)$, we obtain

$$0 = \sup_{\zeta \geq 0} E[V_{t-h}(W_h, \varphi_h, S_h; u)] - V_t(w, \varphi, s; u)$$

$$= \sup_{\zeta \geq 0} E[\int_0^h \left(-\frac{\partial}{\partial t} + \mathcal{L}^\zeta\right) V_r(W_r, \varphi_r, S_r; u) dr]$$

to arrive at

$$(9) \qquad \frac{1}{h} \sup_{\zeta \geq 0} E[\int_0^h \left(-\frac{\partial}{\partial t} + \mathcal{L}^\zeta\right) V_r(W_r, \varphi_r, S_r; u) dr] = 0$$

by virtue of Itô's formula, where

$$\mathcal{L}^\zeta v(t, w, \varphi, s) = \frac{1}{2} \hat{\sigma}(s)^2 \frac{\partial^2}{\partial s^2} v(t, w, \varphi, s) + \hat{b}(s) \frac{\partial}{\partial s} v(t, w, \varphi, s)$$

$$+ \zeta\left(s \frac{\partial}{\partial w} v(t, w, \varphi, s) - \frac{\partial}{\partial \varphi} v(t, w, \varphi, s)\right) - \hat{g}(\zeta) s \frac{\partial}{\partial s} v(t, w, \varphi, s),$$

$$\hat{g}(\zeta) = \gamma g(\zeta) + \int_{(0,\infty)} (1 - e^{-g(\zeta)z}) \nu(dz).$$

We remark that $\mathcal{L}^\zeta$ is of exactly the same form as in [6]. Letting $h \to 0$ in (9), we obtain

$$\frac{\partial}{\partial t} V_r(W_r, \varphi_r, S_r; u) - \sup_{\zeta \geq 0} \mathcal{L}^\zeta V_r(W_r, \varphi_r, S_r; u) = 0.$$

Note that this calculation is nothing but intuitive and formal. However, we can justify the following theorem.

Theorem 4.1. *Assume that h is strictly increasing and $h(\infty) = \infty$. Moreover, assume*

$$(10) \qquad \liminf_{\varepsilon \downarrow 0} \frac{V_t(w, \varphi, s + \varepsilon; u) - V_t(w, \varphi, s; u)}{\varepsilon} > 0$$

for any $t \in (0, 1]$ and $(w, \varphi, s) \in U$. Then, $V_t(w, \varphi, s; u)$ is a viscosity solution of

$$(11) \qquad \frac{\partial}{\partial t} v(t, w, \varphi, s) - \sup_{\zeta \geq 0} \mathcal{L}^\zeta v(t, w, \varphi, s) = 0, \quad (t, w, \varphi, s) \in (0, 1] \times U,$$

where $U = \hat{D} \setminus \partial \hat{D}$ and $\hat{D} = \mathbb{R} \times [0, \infty) \times [0, \infty)$.

Proof. We consider the strategy-restricted version of the value function

$$V_t^L(w, \varphi, s; u) = \sup_{(\zeta_r)_{r \leq t} \in \mathcal{A}_t^L(\varphi)} \mathbb{E}[u(W_t, \varphi_t, S_t)]$$

for $L > 0$, where $\mathcal{A}_t^L(\varphi) = \{(\zeta_r)_{0 \leq r \leq t} \in \mathcal{A}_t(\varphi) ; \sup_{r,\omega} |\zeta_r(\omega)| \leq L\}$. Then, the standard argument (see Section 5.4 in [11]) suggests that $V_t^L(w, \varphi, s; u)$ is a viscosity solution of

$$\frac{\partial}{\partial t} v(t, w, \varphi, s) - \sup_{0 \leq \zeta \leq L} \mathcal{L}^\zeta v(t, w, \varphi, s) = 0, \quad (t, w, \varphi, s) \in (0, 1] \times U$$

because of the compactness of the control region $[0, L]$. Here, we can define $\hat{h}(\zeta)$ as the derivative of $\hat{g}$ by

$$\hat{h}(\zeta) = \left(\gamma + \int_{(0,\infty)} z e^{-g(\zeta)z} v(dz) \right) h(\zeta).$$

Obviously, it holds that $\hat{h}(\infty) = \infty$. Hence, we can apply the same argument as in Section 7.6 in [6] to obtain the assertion by letting $L \to \infty$ and using the stability arguments for viscosity solutions. $\qquad \square$

Since the condition $\lim_{\zeta \to \infty}(h(\zeta)/\zeta) > 0$ implies $\lim_{\zeta \to \infty}(\hat{h}(\zeta)/\zeta) > 0$, we can apply Theorem 5 in [6] to arrive at the following uniqueness theorem.

Theorem 4.2. *Assume that $\hat{\sigma}$ and $\hat{b}$ are both Lipschitz continuous. Assume the hypotheses in Theorem 4.1 and that $\liminf_{\zeta \to \infty}(h(\zeta)/\zeta) > 0$. If a polynomial growth function $v : [0,1] \times \hat{D} \longrightarrow \mathbb{R}$ is a viscosity solution of (11) and satisfies the boundary conditions*

$$
v(0, w, \varphi, s) = u(w, \varphi, s), \ (w, \varphi, s) \in \hat{D},
$$
$$
(12) \quad v(t, w, 0, s) = \mathrm{E}\left[u\left(w, 0, Z\left(t; 0, s\right)\right)\right], \ (t, w, s) \in [0,1] \times \mathbb{R} \times [0, \infty),
$$
$$
v(t, w, \varphi, 0) = u(w, \varphi, 0), \ (t, w, \varphi) \in [0,1] \times \mathbb{R} \times [0, \infty),
$$

then $V_t(w, \varphi, s; u) = v(t, w, \varphi, s)$, where

$$
(13) \qquad Z(t; r, s) = \exp\left(Y\left(t; r, \log s\right)\right) \ (s > 0), \ 0 \ (s = 0).
$$

5. Sell-Off Condition

In this section, we consider the optimal execution problem under the "sell-off condition", which was introduced in [6]. A trader has a certain quantity of shares of a security at the initial time, and must liquidate all of them by the time horizon. Then, the spaces of admissible strategies are reduced to

$$
\mathcal{A}_k^{n,\mathrm{SO}}(\varphi) = \left\{ (\psi_l^n)_l \in \mathcal{A}_k^n(\varphi) \ ; \ \sum_{l=0}^{k-1} \psi_l^n = \varphi \right\},
$$
$$
\mathcal{A}_t^{\mathrm{SO}}(\varphi) = \left\{ (\zeta_r)_r \in \mathcal{A}_t(\varphi) \ ; \ \int_0^t \zeta_r dr = \varphi \right\}.
$$

Now, we define value functions with the sell-off condition by

$$
V_k^{n,\mathrm{SO}}(w, \varphi, s; U) = \sup_{(\psi_l^n)_l \in \mathcal{A}_k^{n,\mathrm{SO}}(\varphi)} \mathrm{E}[U(W_k^n)],
$$
$$
V_t^{\mathrm{SO}}(w, \varphi, s; U) = \sup_{(\zeta_r)_r \in \mathcal{A}_t^{\mathrm{SO}}(\varphi)} \mathrm{E}[U(W_t)]
$$

for a continuous, non-decreasing and polynomial growth function $U : \mathbb{R} \longrightarrow \mathbb{R}$, where $(W_l^n, \varphi_l^n, S_l^n)_{l=0}^k = \Xi_k^n(w, \varphi, s; (\psi_l^n)_l)$ and $(W_r, \varphi_r, S_r)_{0 \le r \le t} = \Xi_t(w, \varphi, s; (\zeta_r)_r)$.

The following theorem is analogous to Theorem 7 in [6].

Theorem 5.1. $V_t^{\mathrm{SO}}(w, \varphi, s; U) = V_t(w, \varphi, s; u)$, *where* $u(w, \varphi, s) = U(w)$.

Proof. The relation $V_t^{\mathrm{SO}}(w, \varphi, s; U) \le V_t(w, \varphi, s; u)$ is trivial, so we show only the assertion $V_t^{\mathrm{SO}}(w, \varphi, s; U) \ge V_t(w, \varphi, s; u)$. Take any $(\zeta_r)_r \in \mathcal{A}_t(\varphi)$ and let $(W_r, \varphi_r, S_r)_r = \Xi_1(w, \varphi, s; (\zeta_r)_r)$. Also, take any $\delta \in (0, t)$. We define an execution strategy $(\zeta_r^\delta)_r \in \mathcal{A}_t^{\mathrm{SO}}(\varphi)$ by $\zeta_r^\delta = \zeta_r \ (r \in [0, t - \delta])$, $\varphi_{t-\delta}/\delta \ (r \in (t - \delta, t])$. Let $(W_r^\delta, \varphi_r^\delta, S_r^\delta)_r = \Xi_t(w, \varphi, s; (\zeta_r^\delta)_r)$. Then, we have $W_{t-\delta} = W_{t-\delta}^\delta \le W_t^\delta$, arriving at $\mathrm{E}[U(W_{t-\delta})] \le \mathrm{E}[U(W_t^\delta)] \le V_t^{\mathrm{SO}}(w, \varphi, s; U)$. Letting $\delta \downarrow 0$, we obtain $\mathrm{E}[U(W_t)] \le$

$V_t^{SO}(w, \varphi, s; U)$ by using the monotone convergence theorem. Since $(\zeta_r)_r \in \mathcal{A}_t(\varphi)$ is arbitrary, we obtain the assertion. $\qquad\square$

By Theorem 5.1, we see that the sell-off condition $\int_0^t \zeta_r dr = \varphi$ does not introduce changes in the (value of the) value function in a continuous-time model. Thus, although the value function in a discrete-time model may depend on whether the sell-off condition is imposed, in the continuous-time model this condition is irrelevant.

The following is also similar to Theorem 7 in [6], which is a version of Theorem 3.1 with the sell-off condition.

Theorem 5.2. *For any* $(w, \varphi, s) \in D$,

$$\lim_{n\to\infty} V_{[nt]}^{n,SO}(w, \varphi, s; U) = V_t^{SO}(w, \varphi, s; U) \ (= V_t(w, \varphi, s; U).)$$

Proof. We may assume $t > 0$. Take any $\delta \in (0, t)$ and let $n > 1/\delta$. Then, each strategy in $\mathcal{A}_{[n(t-\delta)]}^n(\varphi)$ can always be extended to the one in $\mathcal{A}_{[nt]}^{n,SO}(\varphi)$ by liquidating all remaining inventory in the last period. Thus, we see that for $n > 1/\delta$

$$(14) \qquad V_{[n(t-\delta)]}^n(w, \varphi, s; u) \le V_{[nt]}^{n,SO}(w, \varphi, s; U) \le V_{[nt]}^n(w, \varphi, s; u).$$

By Theorem 3.1, we obtain

$$(15) \qquad \lim_{n\to\infty} V_{[n(t-\delta)]}^n(w, \varphi, s; u) = V_{t-\delta}(w, \varphi, s; u),$$

$$\lim_{n\to\infty} V_{[nt]}^n(w, \varphi, s; u) = V_t(w, \varphi, s; u).$$

By (14), (15), and Theorem 3.2, we obtain the assertion. $\qquad\square$

Analogously to Theorem 8 in [6], a result similar to Theorem 3 in [8] holds when $g(\zeta)$ is linear:

Theorem 5.3. *Assume* $g(\zeta) = \alpha_0 \zeta$ *for* $\alpha_0 > 0$.
(i) $V_t^{SO}(w, \varphi, s; U) = \overline{V}_t^\varphi\left(w + \frac{1 - e^{-\gamma\alpha_0\varphi}}{\gamma\alpha_0} s, e^{-\gamma\alpha_0\varphi} s; U\right)$, *where*

$$\overline{V}_t^\varphi(\bar{w}, \bar{s}; U) = \sup_{(\overline{\varphi}_r)_r \in \overline{\mathcal{A}}_t(\varphi)} \mathrm{E}[U(\overline{W}_t)]$$

$$\text{s.t.} \quad d\overline{S}_r = e^{-\gamma\alpha_0\overline{\varphi}_r}\hat{b}(\overline{S}_r e^{\gamma\alpha_0\overline{\varphi}_r})dr + e^{-\gamma\alpha_0\overline{\varphi}_r}\hat{\sigma}(\overline{S}_r e^{\gamma\alpha_0\overline{\varphi}_r})dB_r - \overline{S}_r dG_r,$$

$$d\overline{W}_r = \frac{e^{\gamma\alpha_0\overline{\varphi}_r} - 1}{\gamma\alpha_0}d\overline{S}_r,$$

$$\overline{S}_0 = \bar{s}, \quad \overline{W}_0 = \bar{w}$$

and

$$\overline{\mathcal{A}}_t(\varphi) = \left\{\left(\varphi - \int_0^r \zeta_v dv\right)_{0 \le r \le t} ; \ (\zeta_r)_{0 \le r \le t} \in \mathcal{A}_t^{SO}(\varphi)\right\},$$

where

$$G_r = \int_0^r \int_{(0,\infty)} (1 - e^{-\alpha_0 \zeta_{s-}z}) N(ds, dz).$$

(ii) *If U is concave and $\hat{b}(s) \le 0$ for $s \ge 0$, then*

$$(16) \qquad V_t^{SO}(w, \varphi, s; U) = U\left(w + \frac{1 - e^{-\gamma \alpha_0 \varphi}}{\gamma \alpha_0} s\right).$$

Proof. We can easily confirm assertion (i) by applying Itô's formula to $\overline{S}_r$ and $\overline{W}_r$. By a similar argument to that in Section 7.9 in [6], we obtain

$$E[U(\overline{W}_t)]$$

$$\le U\left(\bar{w} + \int_0^t E\left[\frac{1 - e^{-\gamma \alpha_0 \bar{\varphi}_r}}{\gamma \alpha_0} \hat{b}(\overline{S}_r e^{\gamma \alpha_0 \bar{\varphi}_r}) - \int_{(0,\infty)} \frac{e^{\gamma \alpha_0 \bar{\varphi}_r} - 1}{\gamma \alpha_0} \overline{S}_r (1 - e^{-\alpha_0 \zeta_{r-}z}) v(dz)\right] dr\right)$$

for any $(\bar{\varphi}_r)_r \in \overline{\mathcal{A}}_t(\varphi)$ by virtue of the Jensen inequality. Since $\hat{b}$ is non-positive, the function U is non-decreasing, and the terms

$$1 - e^{-\gamma \alpha_0 \bar{\varphi}_r}, \ e^{\gamma \alpha_0 \bar{\varphi}_r} - 1, \ 1 - e^{-\alpha_0 \zeta_{r-}z}$$

are all non-negative, we see that $E[U(\overline{W}_t)] \le U(\bar{w})$ for any $(\bar{\varphi}_r)_r \in \overline{\mathcal{A}}_t(\varphi)$, which implies $\overline{V}_t^\varphi(\bar{w}, \bar{s}) \le U(\bar{w})$. The opposite inequality $\overline{V}_t^\varphi(\bar{w}, \bar{s}) \ge U(\bar{w})$ is obtained, similarly to the result in Section 7.9 in [6] and the proof of Proposition 11 in [5]. $\square$

We note that the assertion (ii) is the same as Theorem 3 in [8], and in this case we can obtain the explicit form of the value function. The right-hand side of (16) is equal to $Ju(w, \varphi, s)$ for $u(w, \varphi, s) = U(w)$ and the nearly optimal strategy for $V_t^{SO}(w, \varphi, s; U) = V_t(w, \varphi, s; u)$ is given by (8). This implies that whenever considering a linear MI function, a risk-averse (or risk-neutral) trader's optimal strategy of liquidating a security that has a negative risk-adjusted drift is nearly the same as block liquidation (i.e. selling all shares at once) at the initial time.

6. Effect of Uncertainty in MI in the Risk-Neutral Framework

The purpose of this section is to investigate how the noise in the MI function affects the trader. Particularly, we focus on the case where the trader is risk-neutral, i.e. $u(w, \varphi, s) = u_{RN}(w, \varphi, s) = w$.

First, we prepare a value function of the execution problem with a deterministic MI function to perform a comparison with the case of random MI. Let $g_l^n(\psi)$ be as in Section 3, and set $\bar{g}_n(\psi) \equiv E[g_l^n(\psi)] = E[c_l^n] g_n(\psi)$. Note that when $E[c_l^n] = 1$, the function $\bar{g}_n$ is equivalent to g_n. We denote the discrete-time value function by $\overline{V}_k^n(w, \varphi, s; u)$ with an MI function $\bar{g}_n$. Theorem 1 in [6] implies that under

[A] the function $\bar{V}^n_{[nt]}(w, \varphi, s; u)$ converges to the continuous-time value function $\bar{V}_t(w, \varphi, s; u)$, which becomes the same as in (7) by replacing $g(\zeta)$ and L_t with $\tilde{\gamma} g(\zeta)$ and t, i.e., the SDE for $(X_r)_r$ is given as

$$dX_r = \sigma(X_r)dB_r + b(X_r)dr - \tilde{\gamma} g(\zeta_r)dr,$$

where $\tilde{\gamma} = E[L_1]$.

Theorem 6.1. $V^n_k(w, \varphi, s; u_{\mathrm{RN}}) \geq \bar{V}^n_k(w, \varphi, s; u_{\mathrm{RN}})$.

Proof. Take any $(\psi^n_l)_l \in \mathcal{A}^n_k(\varphi)$ and let $(W^n_l, \varphi^n_l, S^n_l)_l = \Xi^n_k(w, \varphi, s; (\psi^n_l)_l)$ be the triplet for $\bar{V}^n_k(w, \varphi, s; u_{\mathrm{RN}})$. Then, the Jensen inequality implies

$$E[W^n_k] = w + \sum_{l=0}^{k-1} E[\psi^n_l S^n_l \exp(-\bar{g}_n(\psi^n_l))]$$

$$= w + \sum_{l=0}^{k-1} E[\psi^n_l S^n_l \exp(-E[c^n_l|\mathcal{F}^n_l]g_n(\psi^n_l))]$$

$$\leq w + \sum_{l=0}^{k-1} E[\psi^n_l S^n_l E[\exp(-c^n_l g_n(\psi^n_l))|\mathcal{F}^n_l]] \leq V^n_k(w, \varphi, s; u_{\mathrm{RN}}).$$

Since $(\psi^n_l)_l$ is arbitrary, we obtain the assertion. □

The above proposition immediately leads us to

$$(17) \qquad V_t(w, \varphi, s; u_{\mathrm{RN}}) \geq \bar{V}_t(w, \varphi, s; u_{\mathrm{RN}}).$$

This inequality shows that noise in MI is welcome since it decreases the liquidation cost for a risk-neutral trader.

For instance, we consider the situation where the trader estimates the MI function from historical data and tries to minimize the expected liquidation cost. Then, a higher sensitivity of the trader to the volatility risk of MI results in a lower estimate for the expected proceeds of the liquidation. This implies that taking into consideration the uncertainty in MI makes the trader prone to underestimating the liquidation cost. Thus, as long as the trader's target is the expected cost, considering the uncertainty in MI is not an incentive for being conservative with respect to the unpredictable liquidity risk. In Section 7, we present the results of numerical experiments conducted in order to simulate above phenomenon.

7. Examples

In this section, we show two examples of our model, both of which are generalizations of the ones in [6].

Motivated by the Black-Scholes type market model, we take $b(x) \equiv -\mu$ and $\sigma(x) \equiv \sigma$ for some constants $\mu, \sigma \geq 0$ and assume $\tilde{\mu} = \mu - \sigma^2/2 > 0$. We also assume that a trader has a risk-neutral utility function $u(w, \varphi, s) = u_{RN}(w) = w$. In this case, if there is no MI, then a risk-neutral trader is afraid of decreasing the expected stock price, and hastens to liquidate the shares.

We consider MI functions which are (log-)linear and (log-)quadratic with respect to liquidation speed, and assume noise distributed by the Gamma distribution.

For the noise part of MI, we set $\gamma_n = \gamma$ for $n \in \mathbb{N}$ and

$$P(c_l^n - \gamma \in dx) = \text{Gamma}(\alpha_1/n, n\beta_1)(dx)$$
$$= \frac{1}{\Gamma(\alpha_1/n)(n\beta_1)^{\alpha_1/n}} x^{\alpha_1/n-1} e^{-x/(n\beta_1)} 1_{(0,\infty)}(x)dx,$$

where $\Gamma(x)$ is the Gamma function. Here, α_1, β_1, and $\gamma > 0$ are constants.

For the deterministic part of MI, we consider two patterns such that $g_n(\psi) = \alpha_0\psi$ and $g_n(\psi) = n\alpha_0\psi^2$ for $\alpha_0 > 0$. In each case, assumptions [A], [B1]–[B3] and [C] are satisfied. The corresponding Lévy measure is

$$\nu(dz) = \frac{\alpha_1}{z} e^{-z/\beta_1} 1_{(0,\infty)}(z)dz.$$

7.1 Log-Linear Impact & Gamma Distribution

Theorem 5.3 directly implies the following theorem.

Theorem 7.1. *It holds that*

$$(18) \qquad V_t(w, \varphi, s; u_{RN}) = w + \frac{1 - e^{-\gamma\alpha_0\varphi}}{\gamma\alpha_0} s$$

for each $t \in (0, 1]$ and $(w, \varphi, s) \in D$.

The implication of this result is the same as in [6]: the right-hand side of (18) is equal to $Ju(w, \varphi, s)$ and converges to $w + \varphi s$ as $\alpha_0 \downarrow 0$ or $\gamma \downarrow 0$, which is the profit gained by choosing the execution strategy of the so-called block liquidation, where a trader sells all shares φ at $t = 0$ when there is no MI. Therefore, the optimal strategy in this case is to liquidate all shares by dividing infinitely within an infinitely short time at $t = 0$ (we refer to such a strategy as a nearly block liquidation at the initial time). Note that the jump part of MI

$$g(\zeta_{r-}) \int_{(0,\infty)} zN(dr, dz)$$

does not influence the value of $V_t(w, \varphi, s; u_{RN})$.

7.2 Log-Quadratic Impact & Gamma Distribution

In [6], we obtained a partial analytical solution to the problem: when φ is sufficiently small or large, we obtain the explicit form of optimal strategies. However, noise in MI complicates the problem, and deriving the explicit solution is more difficult. Thus, we rely on numerical experiments. Owing to the assumption that the trader is risk-neutral, we may assume that an optimal strategy is deterministic. Here, we introduce the following additional condition.

[D] $\gamma \geq \alpha_1 \beta_1 / 8$.

In fact, we can replace our optimization problem with the deterministic control problem

$$f(t, \varphi) = \sup_{(\zeta_r)_r} \int_0^t \exp\left(-\int_0^r q(\zeta_v) dv\right) \zeta_r dr,$$

$$q(\zeta) = \tilde{\mu} + \gamma \alpha_0 \zeta^2 + \alpha_1 \log(\alpha_0 \beta_1 \zeta^2 + 1)$$

for a deterministic process $(\zeta_r)_r$ under the above assumption. We quote Theorem 5 in [5] below.

Theorem 7.2. $V_t(w, \varphi, s; u_{RN}) = w + s f(t, \varphi)$ under [D].

Here, note that Theorems 4.1–4.2 imply the following characterization.

Prop 7.1. $f(t, \varphi)$ is the viscosity solution of

$$(19) \quad \frac{\partial}{\partial t} f - \sup_{\zeta \geq 0}\left\{f + \zeta\left(1 - \frac{\partial}{\partial \varphi} f\right) - \gamma \alpha_0 \zeta^2 f + \alpha_1 \log(\alpha_0 \beta_1 \zeta^2 + 1)\right\} = 0$$

with the boundary condition

$$(20) \quad f(0, \varphi) = \varphi, \quad f(t, 0) = 0.$$

Moreover, if $\tilde{f}$ is a viscosity solution of (19)–(20) and has a polynomial growth rate, then $f = \tilde{f}$.

It is difficult to obtain an explicit form of the solution of (19)–(20). Instead, we solve this problem numerically by considering the deterministic control problem $f_{[nt]}^n(\varphi)$ in the discrete-time model for a sufficiently large n:

$$f_k^n(t, \varphi)$$

$$= \sup_{\substack{(\psi_l^n)_{l=0}^{k-1} \subset [0,\varphi]^k, \\ \sum_l \psi_l^n \leq \varphi}} \sum_{l=0}^{k-1} \psi_l^n \exp\left(-\tilde{\mu} \times \frac{l}{n} - \sum_{m=0}^{l}\left\{n\gamma\alpha_0(\psi_l^n)^2 + \frac{\alpha_1}{n} \log(n^2 \alpha_0 \beta_1 (\psi_l^n)^2 + 1)\right\}\right).$$

We set each parameter as follows: $\alpha_0 = 0.01, t = 1, \tilde{\mu} = 0.05, w = 0, s = 1$ and $n = 500$. For φ, we examine three patterns for $\varphi = 1, 10,$ and 100.

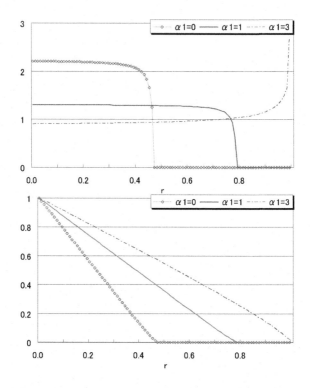

Figure 1. Result for $\varphi = 1$ in the case of fixed γ. Top : The optimal strategy ζ_r. Bottom : The amount of security holdings φ_r.

7.2.1 The case of fixed γ

In this subsection, we set $\gamma = 1$ to examine the effects of the shape parameter α_1 of the noise in MI. Here, we also set $\beta_1 = 2$. As seen in the numerical experiment in [6], the forms of optimal strategies vary according to the value of φ. Therefore, we summarize our results separately for each φ.

Figure 1 shows the graphs of the optimal strategy $(\zeta_r)_r$ and its corresponding process $(\varphi_r)_r$ of the security holdings in the case of $\varphi = 1$, that is, the number of initial shares of the security is small. As found in [6], if there is no noise in the MI function (i.e., if $\alpha_1 = 0$), then the optimal strategy is to sell up the entire amount at the same speed (note that the roundness at the corner in the left graph of Figure 1 represents the discretization error and is not essential). The same tendency is found in the case of $\alpha_1 = 1$, but in this case the execution time is longer than in the case of $\alpha_1 = 0$. When we take $\alpha_1 = 3$, the situation undergoes a complete change. In this case, the optimal strategy is to increase the execution speed as the time horizon approaches.

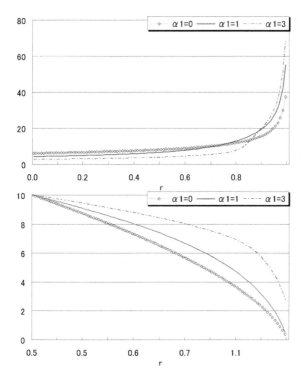

Figure 2. Result for $\varphi = 10$ in the case of fixed γ. Top : The optimal strategy ζ_r. Bottom : The amount of security holdings φ_r.

When the amount of the security holdings is 10, which is larger than in the case of $\varphi = 1$, the optimal strategy and the corresponding process of the security holdings are as shown in Figure 2. In this case, a trader's optimal strategy is to increase the execution speed as the end of the trading time approaches, which is the same as in the case of $\varphi = 1$ with $\alpha_1 = 3$. Clearly, a larger value of α_1 corresponds to a higher speed of execution closer to the time horizon. We should add that a trader cannot complete the liquidation when $\alpha_1 = 3$: However, as mentioned in Section 5, we can choose a nearly optimal strategy from $\mathcal{A}_1^{SO}(\varphi)$ without changing the value of the expected proceeds of liquidation by combining the execution strategy in Figure 2 (with $\alpha_1 = 3$) and the terminal (nearly) block liquidation. See Section 5.2 of [6] for details.

When the amount of the security holdings is too large, as in the case of $\varphi = 100$, a trader cannot complete the liquidation regardless of the value of α_1, as Figure 3 implies. This is similar to the case of $\varphi = 10$ with $\alpha_1 = 3$. The remaining amount of shares of the security at the time horizon is larger for larger noise in MI.

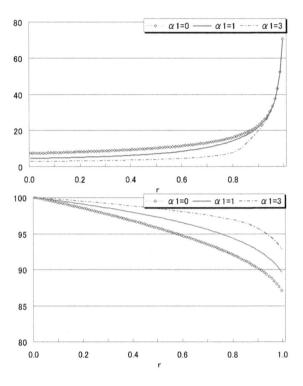

Figure 3. Result for $\varphi = 100$ in the case of fixed γ. Top : The optimal strategy ζ_r. Bottom : The amount of security holdings φ_r.

Note that the trader can also sell all the shares of the security without decreasing the profit by combining the strategy with the terminal (nearly) block liquidation.

7.2.2 The case of fixed $\tilde{\gamma}$

In the above subsection, we presented a numerical experiment performed to compare the effects of the parameter α_1 by fixing γ. Here, we perform numerical comparison from a different viewpoint.

The results in Section 6 imply that taking the uncertainty in MI into account makes a risk-neutral trader optimistic about the estimation of liquidity risks. To obtain a deeper insight, we investigate the structure of the MI function in more detail. In Theorems 3.2(ii) and 7.1, the important parameter is γ, which is the infimum of L_1 and is smaller than (or equal to) $E[L_1]$. We can interpret this as a characteristic feature whereby the (nearly) block liquidation eliminates the effect of positive jumps of $(L_t)_t$. However, there is another decomposition of L_t such that

$$L_t = \tilde{\gamma}t + \int_0^t \int_{(0,\infty)} z\tilde{N}(dr, dz),$$

where

$$\tilde{\gamma} = \gamma + \int_{(0,\infty)} z\nu(dz) = \mathrm{E}[L_1]$$

and

$$\tilde{N}(dr, dz) = N(dr, dz) - \int_{(0,\infty)} z\nu(dz)dr.$$

This representation is essential from the viewpoint of martingale theory. Here, $\tilde{N}(\cdot, \cdot)$ is the compensator of $N(\cdot, \cdot)$ and $\tilde{\gamma}$ can be regarded as the "expectation" of the noise in MI. Just for a risk-neutral world (in which a trader is risk-neutral), as studied in Section 6, we can compare our model with the case of deterministic MI functions as in [6] by setting $\tilde{\gamma} = 1$. Based on this, we conduct another numerical experiment with a constant value of $\tilde{\gamma}$.

Note that in our example

(21) $$\tilde{\gamma} = \gamma + \alpha_1\beta_1$$

and

(22) $$\frac{1}{t}\mathrm{Var}\left(\int_0^t \int_{(0,\infty)} z\tilde{N}(dr, dz)\right) = \alpha_1\beta_1^2$$

hold. Here, (21) (respectively, (22)) corresponds to the mean (respectively, the variance) of the noise in the MI function at unit time. Comparisons in this subsection are performed with the following assumptions. We set the parameters β_1 and γ to satisfy

$$\gamma + \alpha_1\beta_1 = 1, \quad \alpha_1\beta_1^2 = 0.5.$$

We examine the cases of $\alpha_1 = 0.5$ and 1, and compare them with the case of $\gamma = 1$ and $\alpha_1 = 0$.

Figure 4 shows the case of $\varphi = 1$, where the trader has a small amount of security holdings. Compared with the case in Section 7.2.1, the forms of all optimal strategies are the same; that is, the trader should sell the entire amount at the same speed. The execution times for $\alpha_1 > 0$ are somewhat shorter than for $\alpha_1 = 0$.

Figure 5 corresponds to the case of $\varphi = 10$. The forms of the optimal strategies are similar to the case of $\varphi = 10$, $\alpha_1 = 0, 1$ in Section 7.2.1. Clearly, the speed of execution near the time horizon increases with increasing α_1.

The results for $\varphi = 100$ are shown in Figure 6. The forms of the optimal strategies are similar to the case of $\varphi = 100$ in Section 7.2.1. However, in contrast to the results in the previous subsection, the remaining amount of shares of the security at the time horizon is smaller for larger α_1.

Finally, we investigate the total MI cost introduced in [7]:

(23) $$\mathrm{TC}(\Phi_0) = -\log \frac{V_T(0, \Phi_0, s)}{\Phi_0 s}.$$

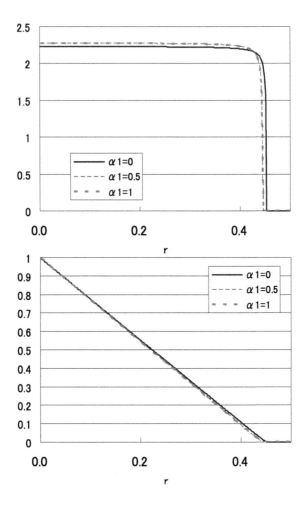

Figure 4. Result for $\varphi = 1$ in the case of fixed $\tilde{\gamma}$. Top : The optimal strategy ζ_r. Bottom : The amount of security holdings φ_r.

When the market is fully liquid and there is no MI, then the total proceeds of liquidating Φ_0 shares of the security are equal to $\Phi_0 s$, because the drift of the price process is negative and the expected price decreases as time passes (the optimal strategy of a risk-neutral trader is block liquidation at the initial time). On the other hand, in the presence of MI, the (optimal) total proceeds decrease to $V_T(0, \Phi_0, s) = \Phi_0 s \times \exp(-\mathrm{TC}(\Phi_0))$. Thus, the total MI cost $\mathrm{TC}(\Phi_0)$ denotes the loss rate caused by MI in a risk-neutral world.

Figure 7 shows the total MI costs in the cases of $\Phi_0 = 1$ and 10. Here, we omit

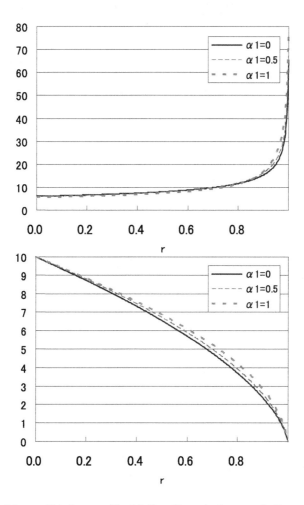

Figure 5. Result for $\varphi = 10$ in the case of fixed $\tilde{\gamma}$. Top : The optimal strategy ζ_r. Bottom : The amount of security holdings φ_r.

the case of $\Phi_0 = 100$ because the amount of shares of the security is too large to complete the liquidation unless otherwise combining terminal block liquidations (which may crash the market). In both cases of $\Phi_0 = 1$ and 10, we find that the total MI cost decreases by increasing α_1. Since the expected value $\tilde{\gamma}$ of the noise in MI is fixed, an increase in α_1 implies a decrease in γ and β_1. Risk-neutral traders seem to be sensitive toward the parameter γ rather than toward α_1, and thus the trader can liquidate the security without concern about the volatility of the noise in MI. Therefore, the total MI cost for $\alpha_1 > 0$ is lower than that for $\alpha_1 = 0$.

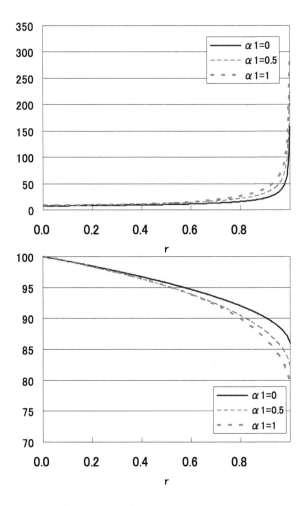

Figure 6. Result for $\varphi = 100$ in the case of fixed $\tilde{\gamma}$. Top : The optimal strategy ζ_r. Bottom : The amount of security holdings φ_r.

8. Concluding Remarks

In this paper, we generalized the framework in [6] and studied an optimal execution problem with random MI. We defined the MI function as a product of an i.i.d. positive random variable and a deterministic function in a discrete-time model. Furthermore, we derived the continuous-time model of an optimization problem as a limit of the discrete-time models, and found out that the noise in MI in the continuous-time model can be described as a Lévy process. Our main results discussed in Sections 3–5 are almost the same as in [6].

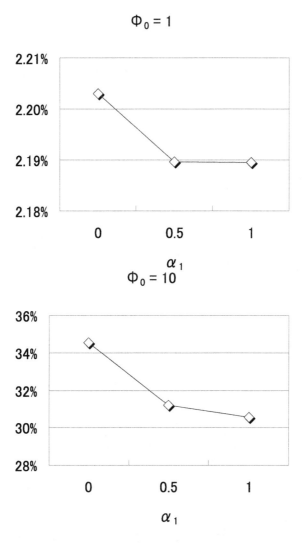

Figure 7. Total MI cost TC(Φ_0) for a risk-neutral trader. Top : the case of $\Phi_0 = 1$. Bottom : the case of $\Phi_0 = 10$. The horizontal axes denote the shape parameter α_1 of the Gamma distribution.

When considering uncertainty in MI, there are two typical barometers of the "level" of MI: γ and $\tilde{\gamma}$. By using the former parameter γ, we can decompose MI into a deterministic part $\gamma g(\zeta_t)dt$ and a pure jump part $g(\zeta_{t-}) \int_{(0,\infty)} zN(dt,dz)$. Then, the pure jump part can be regarded as the difference from the deterministic MI case studied in [6]. On the other hand, as mentioned in Sections 6 and 7, the latter parameter $\tilde{\gamma}$ is important not only in martingale theory but also in a risk-neutral world. Studying $\tilde{\gamma}$ also provides some hints about actual trading practices. Regardless of whether we accommodate uncertainty into MI, it may result in an underestimate of MI for a risk-neutral trader. The risk-neutral setting (i.e. the utility function is set as $u_{RN}(w) = w$) is typical and standard assumption in the study of the execution problem, especially in the limit-order-book model (see, for instance, [1]).

Studying the effects of uncertainty in MI in a risk-averse world is also meaningful. As mentioned in Section 5, when the deterministic part of the MI function is linear, the uncertainty in MI does not significantly influence the trader's behavior, even when the trader is risk-averse. In future work, we will investigate the case of nonlinear MI.

Finally, in our settings, the MI function is stationary in time, but in the real market, the characteristics of MI change according to the time zone. Therefore, it is meaningful to study the case where the MI function is inhomogeneous in time. This is another topic for future work.

References

1. Alfonsi, A., Fruth, A. and Schied, A. (2010), "Optimal execution strategies in limit order books with general shape functions," *Quant. Finance* **10**, 143–157
2. Almgren, R. and N. Chriss (2000), "Optimal execution of portfolio transactions," *J. Risk*, **3**, 5-39.
3. Bertsimas, D. and A. W. Lo (1998), "Optimal control of execution costs," *J. Fin. Markets*, **1**, 1–50.
4. Gatheral, J. (2010), "No-dynamic-arbitrage and market impact," *Quant. Finance* **10**, 749–759
5. Ishitani, K. and T. Kato (2009), "Optimal execution problem with random market impact," *MTEC Journal*, **21**, 83-108. (in Japanese)
6. Kato, T. (2012), "Formulation of an optimal execution problem with market impact: derivation from discrete-time models to continuous-time models," *arXiv preprint*, http://arxiv.org/pdf/0907.3282
7. Kato, T. (2010), "Formulation of an Optimal Execution Problem with Market Impact," Proceedings of The 41th ISCIE International Symposium on Stochastic Systems, Theory and Its Applications (SSS'09), 235–240.
8. Lions, P.-L. and J.-M. Lasry (2007), "Large investor trading impacts on volatility," *Paris-Princeton Lectures on Mathematical Finance 2004, Lecture Notes in Mathematics 1919*, Springer, Berlin, 173-190.
9. Moazeni, S., Thomas F. Coleman and Y. Li (2011), "Optimal execution under jump models for uncertain price impact," to appear in *Journal of Computational Finance*.
10. Subramanian, A. and R. Jarrow (2001), "The liquidity discount," *Math. Finance*, **11**, 447-474.
11. Nagai, H. (1999), *Stochastic Differential Equations*, Kyoritsu-Shuppan (in Japanese).

Optimal Investment Timing and Volume Decisions under Debt Borrowing Constraints [*]

Takashi Shibata[1][†] and Michi Nishihara[2]

[1]Graduate School of Social Sciences, Tokyo Metropolitan University
Email: tshibata@tmu.ac.jp
[2]Graduate School of Economics, Osaka University
Email: nishihara@econ.osaka-u.ac.jp

This paper examines the optimal investment threshold (timing) and volume strategies under debt borrowing constraints on the condition that an increase in investment volume increases cash inflow. As debt borrowing limits increase, the firm is more likely to prefer market debt to bank debt. Although debt borrowing constraints distort the investment threshold, they have no influence on the optimal investment volume. Debt borrowing constraints have the possibility of distorting the interactions between equity values and the investment threshold.

Key words: Investment timing; investment quantity; capital structure, financing constraints; debt structure.

1. Introduction

The purpose of this paper is to consider the interactions between financing and investment decisions under debt borrowing constraints. Modigliani and Miller (1958) find that investment and financing decisions are completely separable in a perfectly competitive market. However, market frictions distort the interactions compared with those in a perfectly competitive market.

---

[*]We would like to thank seminar participants at Akita Prefecture University, Bank of Japan, Chinese University of Hong Kong, Hokkaido University, Hong Kong University of Science and Technology, Kyoto University, Nanzan University, University College London, University of Tokyo, and University of Toronto. This research was partially supported by a Grant-in-Aid from KAKENHI (21241040, 23310103), the Tokyo Metropolitan University, and the Ishii Memorial Securities Research Promotion Foundation.

[†]Corresponding author.

This paper develops a model of financial frictions between investment and financing decisions. To be more precise, we assume that the firm has limited capacity constraints for issuing two kinds of debts: market debt and bank debt. As in Jensen and Meckling (1976), investors are reluctant to lend beyond a certain amount because issuing debt encourages risk shifting from equity holders to debt holders. Following Gertner and Scharfstein (1991), the only difference between bank and market debt in this model is the bankruptcy procedure.[1]

In this paper, we endogenously determine investment (threshold) timing, investment volume, coupon payment, and debt structure (either bank or market debt issuance) under financial frictions. Our model is solved as follows. Given a debt structure, we derive the investment threshold, investment volume, and coupon payment strategies. We then determine the optimal debt structure by comparing the equity values financed by bank debt with those financed by market debt.

Our paper provides several important results. First, as debt borrowing limits increase, the firm is more likely to prefer market debt to bank debt. On the other hand, as volatility increases, the firm is more likely to prefer bank debt to market debt. These results fit well with the empirical findings by Blackwell and Kidwell (1988). Second, the investment thresholds have a U-shaped curve with debt borrowing constraints. These results are the same as in Boyle and Guthrie (2003). Nevertheless, the investment volumes are invariant with respect to debt borrowing constraints. Finally, debt borrowing constraints have the possibility of distorting the symmetric relationship between equity option values and investment thresholds.

The remainder of the paper is organized as follows. Section 2 describes the model, derives the value functions, and formulates our problem. Section 3 reviews two extreme cases of our problem as a benchmark. Section 4 derives our solution numerically and discusses the model's implications. Section 5 presents our concluding remarks.

2. Model

In this section, we begin by setting up the model. We then provide the value functions. Finally, we formulate our model as an investment optimization problem for the constrained firm.

2.1 Setup

A firm possesses the option to invest in a single project at any time. If the investment option is exercised at time t, the firm receives an instantaneous cash inflow δX_t after time t. Here, X_t is given by the following geometric Brownian motion:

(1) $$\mathrm{d}X_t = \mu X_t \mathrm{d}t + \sigma X_t \mathrm{d}z_t^{\mathbb{Q}}, \quad X_0 = x > 0,$$

---

[1] They assume that payments to market lenders cannot be changed outside the formal bankruptcy process and that new owners can recapitalize optimally, although costs are incurred.

where $\mu > 0$ and $\sigma > 0$ are positive constants and z_t^Q denotes a standard Brownian motion defined by a risk-neutral probability space $(\Omega, \mathcal{F}, \mathbb{Q})$. For convergence, we assume $r > \mu$ where $r > 0$ is a risk-free interest rate. Throughout our analysis, it is assumed that the current demand level $X_0 = x$ is sufficiently low that the firm does not invest immediately.

When the firm invests in a project at any time t, it incurs an investment cost, $I(\delta)$ where $I(0) > 0$, $I'(\delta) > 0$, and $I''(\delta) > 0$ for all $\delta > 0$ with $\lim_{\delta \downarrow 0} I'(\delta) = 0$ and $\lim_{\delta \uparrow +\infty} I'(\delta) = +\infty$. At the time of investment, δ is endogenously chosen to maximize the firm's profit.

We assume that the firm issues two classes of perpetual debt: market debt with a promised coupon payment flow c_1 (the subscript "1" indicates market debt financing) and bank debt with a promised coupon payment flow c_2 (the subscript "2" indicates bank debt financing). Following Gertner and Scharfstein (1991) and Bolton and Scharfstein (1996), we assume that the only difference between bank and market debt is the bankruptcy procedure. Now suppose that the firm in financial distress tries to restructure the debt. Under market debt financing, the coupon payments to the market lender cannot be changed outside of the formal bankruptcy process. Under bank debt financing, the coupon payments to the bank lender are reduced in the course of a costless private workout.

Let T_k^i and T_j^d denote the investment (indicated by the superscript "i") and default (indicated by the superscript "d") timings, respectively ($k \in \{0, 1, 2\}, j \in \{1, 2\}$). Note that the investment and default timings depend on the financing method indicated by k, where the subscripts "$k = 0$," "$k = 1$," and "$k = 2$" stand for all-equity financing, market debt financing, and bank debt financing, respectively. Mathematically, the investment and default timings are defined as $T_k^i = \inf\{s \geq 0, X_s \geq x_k^i\}$ and $T_j^d = \inf\{s \geq T_j^i; X_s \leq x_j^d\}$, where x_k^i and x_j^d denote the associated investment and default thresholds, respectively. In particular, x_1^d and x_2^d represent the formal bankruptcy threshold under market debt financing and the coupon switching (negotiation) threshold under bank debt financing, respectively.

The left-hand and right-hand panels of Figure 1 depict the scenarios for market debt financing and bank debt financing, respectively. Recall that the initial value $X_0 = x > 0$ is sufficiently small. In the left-hand panel, once the cash inflow X_t is increased and arrives at the higher boundary x_1^i (called the investment threshold), the firm exercises its investment opportunity by issuing market debt. We call the region after investment Region "a." After investment, if X_t is decreased and hits at the lower boundary x_1^d (called the liquidation threshold), the firm is liquidated (that is, the firm ceases operation) under market debt financing. In the right-hand panel, after investment, if X_t decreases and approaches the lower boundary x_2^d (called the coupon-switching threshold), the coupon payment is reduced by negotiation between the firm and the bank. As in Sundaresan and Wang (2007), the firm and the bank negotiate and divide the surplus by eliminating liquidation based on their relative bargaining powers. Let $\eta \in [0, 1]$ and $1 - \eta$ denote, respectively, the firm's

120

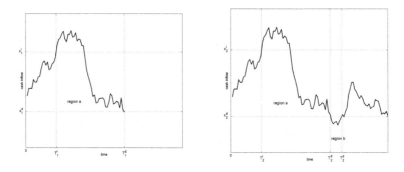

Figure 1. Market debt and bank debt

and bank's bargaining powers. As a result, the normal coupon c_2 is paid in the normal region, while the reduced coupon $s(X_t, \delta_2) < c_2$ is paid in the negotiation region. Since $s(X_t, \delta_2)$ is a linear function of X_t at the equilibrium,[2] the firm does not consider liquidation under bank debt financing. Furthermore, if X_t arrives at x_2^d from the below, the reduced coupon $s(X_t, \delta_2)$ is changed to the normal coupon c_2. Thus, x_2^d is the coupon switching threshold. After investment, the regions $\{X_t \geq x_2^d\}$ and $\{X_t < x_2^d\}$ are called Regions "a" and "b," respectively.

In the next subsection, we derive the value functions after investment.

2.2 Value function for all-equity financed firm

In this subsection, we derive the equity value function for all-equity-financed firm after investment. Let $E_0^a(X_t, \delta_0)$ denote the equity value at any time t after investment under all-equity financing, where the superscript "a" represents the value after investment.

Now we assume $t > T_0^i$, which means that the investment is exercised. The equity value, $E_0^a(X_t, \delta_0)$, is given by

(2) $$E_0^a(X_t, \delta_0) = \mathbb{E}_t^Q\left[\int_t^{+\infty} e^{-r(u-t)}(1-\tau)\delta_0 X_u du \right] = \Pi \delta_0 X_t,$$

where $\tau > 0$ represents the corporate tax, $\mathbb{E}_t^Q$ denotes the expectation operator at time t under probability measure Q, and $\Pi := (1-\tau)/(r-\mu) > 0$.

2.3 Value functions for market debt financed firm

This subsection provides the value functions after market debt is issued at the time of investment. Let us denote by $E_1^a(X_t, c_1, \delta_1)$ and $D_1^a(X_t, c_1, \delta_1)$ the respective equity and debt values for the firm financed by bank debt.

---

[2]See Sundaresan and Wang (2007) for details.

The equity value, $E_1^a(X_t, c_1, \delta_1)$, is defined by

$$(3) \qquad E_1^a(X_t, c_1, \delta_1) = \sup_{T_1^d} \mathbb{E}_t^Q \left[\int_t^{T_1^d} e^{-r(u-t)}(1-\tau)(\delta_1 X_u - c_1) du \right].$$

Using standard arguments, this value is rewritten as

$$(4) \quad E_1^a(X_t, c_1, \delta_1) = \max_{x_1^d} \Pi \delta_1 X_t - (1-\tau)\frac{c_1}{r} - \left(\delta_1 \Pi x_1^d - (1-\tau)\frac{c_1}{r} \right)\left(\frac{X_t}{x_1^d}\right)^\gamma,$$

where $\gamma := 1/2 - \mu/\sigma^2 - \sqrt{(\mu/\sigma^2 - 1/2)^2 + 2r/\sigma^2} < 0$. Then, as in Black and Cox (1976), the liquidation (formal bankruptcy) threshold, $x_1^d(\delta_1, c_1)$, is given by

$$(5) \qquad x_1^d(c_1, \delta_1) = \underset{x_1^d}{\arg\max}\, E_1^a(X_t, c_1, \delta_1) = \kappa_1^{-1}\frac{c_1}{\delta_1},$$

where $\kappa_1 := (\gamma - 1)\Pi r/(\gamma(1-\tau)) \geq 0$. Note that $\lim_{c_1 \downarrow 0} x_1^d(c_1, \delta_1) = 0$. The debt value, $D_1^a(X_t, c_1, \delta_1)$, is given by

$$(6) \quad D_1^a(X_t, c_1, \delta_1) = \mathbb{E}_t^Q\left[\int_t^{T_1^d} e^{-r(u-t)}c_1 du + e^{-r(T_1^d - t)}(1-\alpha)\Pi\delta_1 x_1^d(\delta_1, c_1) \right]$$

$$= \frac{c_1}{r}\left(1 - \left(\frac{\kappa_1 \delta_1 X_t}{c_1}\right)^\gamma \right) + (1-\alpha)\Pi\kappa_1^{-1}c_1\left(\frac{\kappa_1 \delta_1 X_t}{c_1}\right)^\gamma,$$

where $\alpha \in (0, 1)$ is the proportional formal bankruptcy cost. The total firm value, $V_1^a(X_t, c_1, \delta_1) = E_1^a(X_t, c_1, \delta_1) + D_1^a(X_t, c_1, \delta_1)$, is equal to

$$(7) \qquad V_1^a(X_t, c_1, \delta_1) = \Pi\delta_1 X_t + \tau\frac{c_1}{r}\left(1 - \left(\frac{\kappa_1 \delta_1 X_t}{c_1}\right)^\gamma \right) - \alpha\Pi\kappa_1^{-1}c_1\left(\frac{\kappa_1 \delta_1 X_t}{c_1}\right)^\gamma.$$

2.4 Value functions for bank debt financed firm

This subsection provides the value functions after bank debt is issued at the time of investment. Recall that there are two types of region: normal and coupon reduction. Let $E_2^a(X_t, c_2, \delta_2)$ and $E_2^b(X_t, c_2, \delta_2)$ denote the equity values in the normal and coupon reduction (negotiation) regions, respectively, where the superscripts "a" and "b" indicate the normal and coupon reduction regions, respectively.

The equity values in the normal and coupon reduction regions after investment are given by

$$(8) \quad E_2^a(X_t, c_2, \delta_2)$$

$$= \sup_{T_2^d \geq t > T_2^i} \mathbb{E}_t^Q\left[\int_t^{T_2^d} e^{-r(u-t)}(1-\tau)(\delta_2 X_u - c_2) du + e^{-r(T_2^d - t)}E_2^b(X_{T_2^d}, c_2, \delta_2) \right]$$

and

(9) $E_2^b(X_t, c_2, \delta_2)$

$$= \mathbb{E}_t^Q\left[\int_t^{T_2^d} e^{-r(u-t)}(1 - \tau)(\delta_2 X_u - s(X_u, \delta_2))du + e^{-r(T_2^d-t)}E_2^a(X_{T_2^d}, c_2, \delta_2)\right].$$

Note that the coupon payments in (8) and (9) are defined by c_2 and $s(X_t, \delta_2)$, respectively. As shown in Sundaresan and Wang (2007), the reduced coupon payment is obtained by

(10) $$s(x, \delta_2) = (1 - \alpha\eta)(1 - \tau)\delta_2 x.$$

The equity values are rewritten as

(11) $E_2^a(X_t, c_2, \delta_2)$

$$= \max_{x_2^d} \Pi\delta_2 X_t - (1 - \tau)\frac{c_2}{r} - \left\{(1 - \alpha\eta)\Pi\delta_2 x_2^d - \frac{c_2}{r}\left(1 - \tau - \tau\frac{\eta\gamma}{\beta - \gamma}\right)\right\}\left(\frac{X_t}{x_2^d}\right)^\gamma,$$

where $\beta := 1/2 - \mu/\sigma^2 + \sqrt{(\mu/\sigma^2 - 1/2)^2 + 2r/\sigma^2} > 1$ and

(12) $$E_2^b(X_t, c_2, \delta_2) = \eta\left\{\alpha\Pi\delta_2 X_t - \frac{\tau c_2}{r}\frac{\gamma}{\beta - \gamma}\left(\frac{X_t}{x_2^d}\right)^\beta\right\}.$$

Note that (11) and (12) are defined by the regions of $\{X_t \geq x_2^d\}$ and $\{X_t < x_2^d\}$, respectively. Similar to (5), the coupon switching threshold is chosen to satisfy $\partial E_2^a/\partial x_2^{i*} = \partial E_2^b/x_2^{i*}$, i.e.,

(13) $$x_2^d(c_2, \delta_2) = \kappa_2^{-1}\frac{c_2}{\delta_2},$$

where $\kappa_2 := (\gamma - 1)(1 - \alpha\eta)\Pi r/(\gamma(1 - \tau(1 - \eta)) > 0$. Note that $x_2^d(c_2, \delta_2)$ is a linear function of c_2 with $\lim_{c_2 \downarrow 0} x_2^d(c_2, \delta_2) = 0$.

The debt values in the normal and coupon reduction regions, $D_2^a(X_t, c_2, \delta_2)$ and $D_2^b(X_t, c_2, \delta_2)$, are defined by

(14) $D_2^a(X_t, c_2, \delta_2) = \mathbb{E}_t^Q\left[\int_t^{T_2^d} e^{-r(u-t)}c_2 du + e^{-r(T_2^d-t)}D_2^b(X_{T_2^d}, c_2, \delta_2)\right],$

$$= \frac{c_2}{r} + (1 - \alpha\eta)\Pi\delta_2 x_2^d(c_2, \delta_2)\left(\frac{X_t}{x_2^d(c_2, \delta_2)}\right)^\gamma$$

$$- \frac{c_2}{r}\left(1 - \tau + \tau\frac{\beta}{\beta - \gamma} - \tau\frac{\eta\gamma}{\beta - \gamma}\right)\left(\frac{X_t}{x_2^d(c_2, \delta_2)}\right)^\gamma$$

and

$$(15)\ D_2^b(X_t, c_2, \delta_2) = \mathbb{E}_t^Q\left[\int_t^{T_2^d} e^{-r(u-t)} s(X_u, \delta_2) du + e^{-r(T_2^d-t)} D_2^a(X_{T_2^d}, c_2, \delta_2)\right]$$

$$= (1 - \alpha\eta)\Pi\delta_2 X_t - (1 - \eta)\frac{\tau c_2}{r}\frac{\gamma}{\beta - \gamma}\left(\frac{X_t}{x_2^d(c_2, \delta_2)}\right)^\beta,$$

respectively.

The total firm value in the normal region, $V_2^a(X_t, c_2, \delta_2) = E_2^a(X_t, c_2, \delta_2) + D_2^a(X_t, c_2, \delta_2)$, is equal to

$$(16)\qquad V_2^a(X_t, c_2, \delta_2) = \Pi\delta_2 X_t + \tau\frac{c_2}{r}\left(1 - \frac{\beta}{\beta - \gamma}\left(\frac{X_t}{x_2^d(c_2, \delta_2)}\right)^\gamma\right).$$

Note that there is no bankruptcy cost term in (16). We also define the total firm value in the negotiation region as $V_2^b(X_t, c_2, \delta_2) := E_2^b(X_t, c_2, \delta_2) + D_2^b(X_t, c_2, \delta_2)$.

2.5 Problem for the constrained levered firm

In this subsection, we formulate the optimization problem for the constrained levered firm.

Suppose that the debt structure j is given ($j \in \{1, 2\}$). The firm's debt borrowing amount is constrained, i.e.,

$$(17)\qquad \frac{D_j^a(x_j^i, c_j, \delta_j)}{I(\delta_j)} \leq q,$$

for some constant $q \geq 0$. This implies that the ratio of the debt borrowing amount to the investment cost is restricted by the ratio q.

Let $E_j^{**}(x)$ denote the equity option value of the constrained levered firm before investment for a given j ($j \in \{1, 2\}$), where the superscript "$**$" represents the optimum for the constrained problem. Given a debt structure j, the equity option value is defined as

$$(18)\quad E_j^{**}(x) := \sup_{T_j^i, c_j, \delta_j} \mathbb{E}_0^Q\left[e^{-rT_j^i}\left\{E_j^a(X_{T_j^i}, c_j, \delta_j) - \left(I(\delta_j) - D^a(X_{T_j^i}, c_j, \delta_j)\right)\right\}\right]$$

subject to $D_j^a(X_{T_j^i}, c_j, \delta_j)/I(\delta_j) \leq q$. Using the standard arguments in Dixit and Pindyck (1994), we have $\mathbb{E}_0^Q[e^{-rT_j^i}] = (x/x_j^i)^\beta$, where $x_j^i > X_0 = x > 0$.

The optimization problem for the constrained firm is formulated as

$$(19)\qquad E^{**}(x) = \max\{E_1^{**}(x), E_2^{**}(x)\},$$

where $x < \min\{x_1^i, x_2^i\}$ and

$$(20) \qquad E_j^{**}(x) = \max_{x_j^i, c_j, \delta_j} \left(\frac{x}{x_j^i}\right)^{\beta} \{V_j^a(x_j^i, c_j, \delta_j) - I(\delta_j)\}$$

$$(21) \qquad \text{subject to} \quad \frac{D_j^a(x_j^i, c_j, \delta_j)}{I(\delta_j)} \leq q.$$

3. Some preliminaries

In this section, we review two extreme cases briefly before analyzing the optimal investment strategies for a constrained levered firm. The first is the optimization problem for the unlevered (all-equity financed) firm ($q \downarrow 0$). The second is the problem for the non-constrained levered firm ($q \uparrow +\infty$).

3.1 Problem for the unlevered firm

In this subsection, we assume that $q = 0$, i.e., $D_j^a(x_j^i, c_j, \delta_j) = 0$ for all j ($j \in \{1, 2\}$). This problem is equivalent to the simple version of the seminal model by McDonald and Siegel (1986).

Let us denote by $E_0^*(x)$ the equity option value for the unlevered (all-equity financed) firm before investment (the superscript "$*$" represents the optimum for the non-constrained problem). The optimization problem for the all-equity financed firm is given as

$$(22) \qquad E_0^*(x) = \max_{x_0^i, \delta_0} \left(\frac{x}{x_0^i}\right)^{\beta} \{\Pi\delta_0 x_0^i - I(\delta_0)\},$$

where $x < x_0^i$. Recall that $E_0^a(x_0^i, \delta_0) = E_j^a(x_0^i, 0, \delta_0) = V_j^a(x_0^i, 0, \delta_0) = \Pi\delta_0 x_0^i$.

The optimal investment volume δ_0^* is given by δ^* satisfying

$$(23) \qquad \theta := \frac{\beta}{\beta - 1} = \frac{\delta^* I'(\delta^*)}{I(\delta^*)}.$$

The investment threshold under all-equity financing is obtained as

$$(24) \qquad x_0^{i*} = \frac{\theta}{\Pi} \frac{I(\delta^*)}{\delta^*}.$$

Substituting δ^* in (23) and x_0^{i*} in (24) into (22) gives

$$(25) \qquad E_0^*(x) = \left(\frac{x}{x_0^{i*}}\right)^{\beta} (\theta - 1)I(\delta^*).$$

We use these solutions and value as a benchmark.

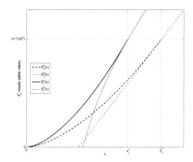

Figure 2. Equity option value $E_k^*(x)$ ($k \in \{0, 1, 2\}, j \in \{1, 2\}$)

3.2 Problem for the non-constrained levered firm

In this subsection, we assume that q is sufficiently large ($q \uparrow +\infty$). This problem is regarded as the extended model of Sundaresan and Wang (2007). Let $E_j^*(x)$ denote the equity option value of the non-constrained levered firm before investment, given a debt structure j ($j \in \{1, 2\}$). The problem for the non-constrained levered firm is formulated as (19) and (20) by removing (21) where the superscript "**" is changed to "*."

Given a debt structure j ($j \in \{1, 2\}$), the solutions and value are as follows.

$$(26) \qquad \delta_j^* = \delta^*, \quad x_j^{i*} = \psi_j x_0^{i*}, \quad c_j^* = \frac{\kappa_j}{h_j} \delta^* x_j^{i*}, \quad x_j^{d*} = \frac{1}{h_j} x_j^{i*},$$

where

$$(27) \qquad \begin{aligned} h_1 &:= \left(1 - \gamma\left(1 + \alpha \tfrac{1-\tau}{\tau}\right)\right)^{-1/\gamma} \geq 1, & \psi_1 &:= \left(1 + \tfrac{\tau}{1-\tau} \tfrac{1}{h_1}\right)^{-1} \leq 1, \\ h_2 &:= \left(\tfrac{\beta}{\beta-\gamma}(1-\gamma)\right)^{-1/\gamma} \geq 1, & \psi_2 &:= \left(1 + \tfrac{\tau(1-\alpha\eta)}{1-\tau(1-\eta)} \tfrac{1}{h_2}\right)^{-1} \leq 1. \end{aligned}$$

The equity option value before investment is

$$(28) \qquad E_j^*(x) = \psi_j^{-\beta} E_0^*(x) = \left(\frac{x}{x_j^{i*}}\right)^\beta (\theta - 1) I(\delta^*),$$

where $x < x_j^{i*}$. Based on the solutions and value for a given debt structure j, we have the following results.

Lemma 3.1. *Consider the optimization problem for the non-constrained levered firm. First, we obtain $\delta^* = \delta_j^*$ for any j ($j \in \{1, 2\}$). Second, we have $E_j^*(x_j^{i*}) = E_0^*(x_0^{i*}) = (\theta - 1)I(\delta^*)$ for any j. Third, we have $x_j^{i*} \leq x_0^{i*}$ if and only if $E_j^*(x) \geq$*

$E_0^*(x)$ for any j. Finally, if $\psi_j \leq \psi_h$, we obtain $E^*(x) = E_j^*(x) \geq E_h^*(x)$ and $x_j^{i*} \leq x_h^{i*}$ for any j and h ($j, h \in \{1, 2\}, j \neq h$).

The third result is a "leverage effect."[3] In the final statement, the equity value is larger if and only if its investment threshold is smaller. Thus, we call the final results a "symmetric relationship" between thresholds and values.

Figure 2 depicts the equity option value $E_k^*(x)$ with the state variable x ($k \in \{0, 1, 2\}$). We see that $x_j^{i*} < x_0^{i*}$, $E_j^*(x) > E_0^*(x)$ where $x < x_j^{i*}$, and $E_j^*(x_j^{i*}) = E_0^*(x_0^{i*})$ for any j ($j \in \{1, 2\}$). We also confirm that $E_j^*(x) \geq E_h^*(x)$ if and only if $x_j^{i*} \leq x_h^{i*}$ for j and h ($j, h \in \{1, 2\}, j \neq h$).

4. Main results

This section provides the solution to the problem formulated in (19). We cannot obtain the analytical solution. In this paper, we consider the solution numerically.

Before solving the solution, we derive the critical point in order to recognize whether the financing constraint is binding or not. We have the following result.

Lemma 4.1. *Given a debt structure j ($j \in \{1, 2\}$), there exists a unique x_j^p satisfying*

$$(29) \qquad \frac{D_j^a(x_j^p, c_j(x_j^p, \delta^*), \delta^*)}{I(\delta^*)} = q,$$

where $c_j(x, \delta^*)$ is given as[4]

$$(30) \qquad c_j(x, \delta^*) = \frac{\kappa_j}{h_j} \delta^* x.$$

Then, if $x_j^{i*} > x_j^p$, the firm is financially constrained. Otherwise, it is not.

We show the proof. First, the left-hand side of (29), $D_j^a(x, c_j(x, \delta^*), \delta^*)/I(\delta^*)$, is a strictly monotonically increasing continuous function of x with $\lim_{x \downarrow 0} D_j^a(x, c_j(x, \delta^*), \delta^*)/I(\delta^*) = 0$ and $\lim_{x \uparrow +\infty} D_j^a(x, c_j(x, \delta^*), \delta^*)/I(\delta^*) = +\infty$ for any j ($j \in \{1, 2\}$). It is straightforward to determine that there exists a unique value x_j^p. Second, suppose $x_j^{i*} \geq x_j^p$. Then the firm would prefer to borrow more than the amount $qI(\delta^*)$ to maximize the total firm value. However, the maximum borrowing amount is $qI(\delta^*)$. Thus, the firm is financially constrained by its debt borrowing limit. Suppose instead that $x_j^{i*} < x_j^p$. Then, the firm is not constrained. We complete the proof.

---

[3] See Myers (1977) for details.

[4] This result is obtained by Leland (1994). Note that we have $c_j^* = c_j(x_j^{i*}, \delta^*)$ in (26).

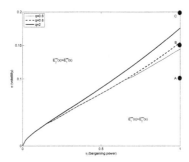

Figure 3. Regions of $E_j^{**}(x) \geq E_h^{**}(x)$ in (η, σ) space $(j, h \in \{1, 2\}, j \neq h)$

In order to provide the numerical solution to the problem defined by (19), the investment cost function is assumed to be

$$(31) \qquad I(\delta_j) = \delta_0 + \delta_j^2, \quad \delta_0 > 0, \quad j \in \{1, 2\}.$$

Suppose that the basic parameters are $r = 0.09$, $\mu = 0.02$, $\delta_0 = 4$, $\tau = 0.15$, $\alpha = 0.4$, and $x = 0.4$.[5]

Figure 3 depicts the regions of $E_j^{**}(x) > E_h^{**}(x)$ in (η, σ) space $(j, h \in \{1, 2\}, j \neq h)$. The three lines indicate the boundaries of $E_1^{**}(x) = E_2^{**}(x)$ for $q = 2$, $q = 0.8$, and $q = 0.6$. Under the basic parameters, the boundary for $q = 2$ is the same as the one for $q \uparrow +\infty$ (i.e., the one for the non-constrained levered firm). For regions with smaller σ and larger η, we see $E_1^{**}(x) \geq E_2^{**}(x)$, i.e., the firm prefers market debt financing. For regions with larger σ and smaller η, in contrast, we have $E_2^{**}(x) \geq E_1^{**}(x)$, i.e., the firm prefers bank debt financing. Importantly, an increase in q enlarges the regions of $E_1^{**}(x) > E_2^{**}(x)$. We summarize these results as follows.

Observation 1. Consider the optimization problem for the constrained levered firm defined by (19). Whether bank or market debt is issued depends on the combination of the three key parameters: q (debt issuance friction), η (bargaining power), and σ (cash inflow volatility). As debt borrowing capacity increases, the firm becomes more likely to issue market debt.

The results obtained by Observation 1 are mostly related to the stylized facts. According to the definition by Rajan (1992), firms with larger q and smaller σ approximate large/mature corporations. Based on this definition, large/mature

---

[5]Under the basic parameters, $x = 0.4$ always satisfies $x < \min\{x_j^{i*}, x_j^{i**}\} < x_0^{i*}$ for all j ($j \in \{1, 2\}$).

128

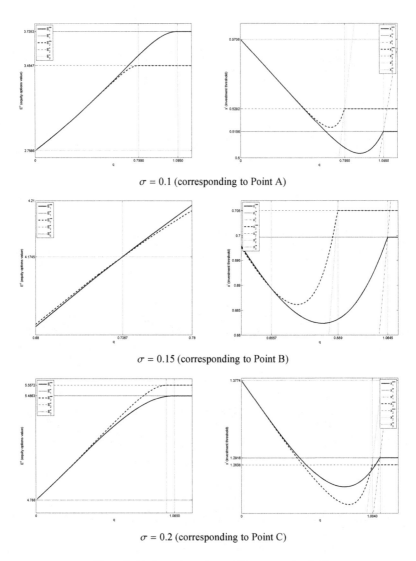

$\sigma = 0.1$ (corresponding to Point A)

$\sigma = 0.15$ (corresponding to Point B)

$\sigma = 0.2$ (corresponding to Point C)

Figure 4. Equity option values and investment thresholds

(small/young) firms are more likely to issue market (bank) debt. These results fit well with the empirical findings by Blackwell and Kidwell (1988).

Figure 4 shows the equity option values and the investment thresholds with q. The parameters of the top, middle, and bottom panels correspond to those of

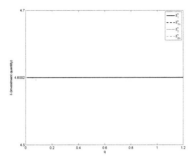

Figure 5. Invariance of debt borrowing amount to investment volume

Points A, B, and C of Figure 3, respectively.

The three left-hand panels of Figure 4 demonstrate the equity option values before investment with q. It is clear that $E_j^{**}(x)$ is monotonically increasing with q for any j ($j \in \{1,2\}$). In the top-left panel, the firm prefers market debt to bank debt for all q. In the middle-left panel, the firm prefers bank debt for $q < \hat{q} = 0.7357$, but market debt for $q \geq \hat{q}$. In the bottom-left panel, the firm prefers bank debt for all q. We confirm that these results correspond to those in Figure 3.

The three right-hand panels of Figure 4 depict the investment thresholds with q. We see that x_k^{i*} does not depend on q for any k ($k \in \{0,1,2\}$). If $x_j^{i*} > x_j^{p}$ for any j ($j \in \{1,2\}$), the firm is financially constrained by the debt borrowing limit. We see that x_j^{i**} is non-monotonic with q. The next observation summarizes the property of investment thresholds under debt borrowing constraints.

Observation 2. Consider the optimization problem for the constrained levered firm defined by (19). We have $x_j^{i**} \leq x_0^{i*}$ for any j ($j \in \{1,2\}$). The constrained investment thresholds x_j^{i**} have a U-shaped curve with q.

The first statement supports the existence of a leverage effect of debt even though the debt borrowing amount is limited. The second statement implies that x_j^{i**} is not always higher than x_j^{i*}. This result is the same as in Shibata and Nishihara (2012, 2013).[6]

We consider the optimal investment volumes with q. The parameters are $\eta = 1$ and $\sigma = 0.15$. The other parameters are the same as in the previous numerical examples. In Figure 5, we see that $\delta^* = \delta_j^{**}$ is constant with respect to q ($j \in \{1,2\}$).[7] Consequently, we have the following result.

---

[6]See Shibata and Nishihara (2012) for the intuition of the U-shaped curve in detail.

[7]We show the analytical proof of this result. However, we omit the proof in this paper.

Observation 3. Consider the optimization problem for the constrained levered firm defined by (19). The investment volumes are invariant with debt borrowing limits.

Observation 3 implies that the optimal investment volumes under market debt are the same as those under bank debt. Wong (2010) has already shown an invariance of δ_1^{**} with q. The new result is that we find an invariance of δ_2^{**} with q.

As in Lemma 3.1, we have $E_j^*(x) \geq E_h^*(x)$ if and only if $x_j^{i*} \leq x_h^{i*}$ when $q \uparrow +\infty$ $(j, h \in \{1, 2\}, j \neq h)$. That is, under no financing constraints, the investment thresholds under market debt are smaller than those under bank debt whenever market debt is preferred to bank debt. In the two bottom panels, on the other hand, we see that $E_1^{**}(x) > E_2^{**}(x)$ and $x_1^{i**} > x_2^{i**}$ for the regions close to $q = 1.0040$. This implies that the investment thresholds under market debt are larger than those under bank debt even though market debt is preferred to bank debt. These results provide the following observation.

Observation 4. Consider the optimization problem for the constrained levered firm defined by (19). When financing constraints are effective, we do not always have a symmetric relationship between equity option values and investment thresholds.

5. Concluding remarks

This paper examines investment timing and volume under debt borrowing constraints. An increase in debt borrowing constraints significantly enlarges the set of market debt preference schemes. Consequently, debt borrowing constraints distort the investment timing, coupon payments. Nevertheless, they do not lead to investment volume efficiency. Debt borrowing constraints have the possibility of breaking down the symmetric relationship between equity option values and investment thresholds. As a result, we are able to discuss the effects of financing constraints.

Some extensions of the model would prove interesting. For example, investors might employ a constraint depending on firm's financial affairs. It might be interesting to investigate whether such a financing constraint maintains a U-shaped relationship between investment timing and market friction.

References

1. Black, F. and Cox, J. C. (1976), "Valuing corporate securities: Some effects of bond indenture provisions." *Journal of Finance* 31, 351–367.
2. Blackwell, D. W. and Kidwell, D. S. (1988), "An investigation of cost differences between public sales and public placements of debt." *Journal of Financial Economics* 22, 253–278.
3. Bolton, P. and Scharfstein, D. S. (1996), "Optimal debt structure and the number of creditors," *Journal of Political Economy* 104, 1–25.

4. Boyle, G. W. and Guthrie, G. A. (2003), "Investment, uncertainty, and liquidity," *Journal of Finance* 63, 2143–2166.

5. Dixit, A. K. and Pindyck, R. S. (1994), *Investment under uncertainty*, Princeton University Press.

6. Gertner, R. H. and Scharfstein, D. S. (1991), "A theory of workouts and the effect of reorganization law," *Journal of Finance* 46, 1189–1222.

7. Jensen, M. C. and Meckling, W. H. (1976), "Theory of the firm: Managerial behavior, agency costs and ownership structure," *Journal of Financial Economics* 3, 305–360.

8. Leland, H. E. (1994), "Corporate debt value, bond covenants, and optimal capital structure," *Journal of Finance* 49, 1213–1252.

9. Modigliani, F. and Miller, M. H. (1958), "The cost of capital, corporate finance, and the theory of investment," *American Economic Review* 48, 261–297.

10. McDonald, R. and Siegel, D. R. (1986), "The value of waiting to invest," *Quarterly Journal of Economics* 101, 707–727.

11. Myers, S. C. (1977), "Determinants of corporate borrowing," *Journal of Financial Economics* 5, 147–175.

12. Rajan, R. (1992), "Insiders and outsiders: The choice between informed and arm's length debt." *Journal of Finance* 47, 1367–1400.

13. Shibata, T. and Nishihara, M. (2012), "Investment timing under debt issuance constraint," *Journal of Banking and Finance* 36, 981–991.

14. Shibata, T. and Nishihara, M. (2013), "Investment timing, debt structure, and financing constraints," *Working Paper*, Tokyo Metropolitan University.

15. Sundaresan, S. and Wang, N. (2007), "Investment under uncertainty with strategic debt service," *American Economic Review Paper and Proceedings* 97, 256–261.

16. Wong, K. P. (2010), "On the neutrality of debt in investment intensity," *Annal of Finance* 6, 335–356.

Fractional Brownian Motions in Financial Models and Their Monte Carlo Simulation

Chun Ming Tam

Graduate School of Social Sciences, Tokyo Metropolitan University
Email: tam-chunmingjeffy@ed.tmu.ac.jp

Fractional Brownian motion (fBm) was first introduced within a Hilbert space framework by Kolmogorov [1], and further studied and coined the name 'fractional Brownian motion' in the 1968 paper by Mandelbrotand Van Ness [2].

In recent years, it has been steadily gaining attention in the area of finance, as it appears that the traditional stochastic volatility model driven by ordinary Brownian motion implies geometric time-decay of the volatility smile, which is much faster than real market data. Such feature is called volatility persistence, and has been largely ignored because of the difficulty to capture it within the ordinary stochastic volatility framework. Several modeling approaches have been suggested capturing this persistence either via a unit-root or long memory process. To keep the pricing framework intact, it is more interesting to study the long memory process, and fBm is a particular good match due to its similarity to the ordinary Brownian motion and its Gaussian properties.

In this paper, we will outline several approaches to simulate fractional Brownian motion with $H > 1 = 2$, where H is a parameter used to generalize Brownian motion into fractional Brownian motion. Through Monte-Carlo simulation, the implied volatility surface is constructed, one of the main result in the research is that, it is discovered that imposing correlation between the fractional Brownian motion driven stochastic volatility and the ordinary Brownian motion driven asset process does not translates into skewness of the implied volatility surface, this is further supported by E.Alos's paper [3] on the observation of the close-to-maturity volatility surface. Unfortunately, since explicit pricing of the Europen option under the fractionally-driven stochastic volatility is not available in closed-form, and market participants have to rely on computational intensive method such as Monte-Carlo simulation, this motivated me to explore an approximation approach proposed by my colleague, Hideharu Funahashi [4], as a starting point in order to come up with a rather robust simulation approach, reducing the brute-force simulation dimension from three to just one, rendering it a computationally inexpensive pricing scheme. As mentioned before, the skewness cannot be modeled by simply imposing correlation; it is necessary to impose an ordinary Brownian motion in the stochastic volatility as well. Our approximation-based-simulation scheme is also capable of pricing European option under this rather complicated stochastic environment, which captures the skewness, smile and persistence of the implied volatility surface.

Key words: Fractional Brownian Motion, Stochastic Volatility, Asymptotic Approximation, Implied Volatility Surface, Volatility Persistence.

1. Financial Motivation and Backgrounds

1.1 Motivation

Numerous empirical studies have pointed out that, in options markets, the implied volatility back-out from the Black-Scholes equation displays volatility skews or smiles; the smile effect, which is well known to practitioners, refers to the U-shape price distortion on the implied volatility surface.

In Hull and White [3] and Scott [4], they have proposed this feature of volatility to be captured by stochastic regime, known as the stochastic volatility model:

(1.1)
$$\begin{cases} \frac{dS(t)}{S(t)} = \mu(t, S_t)dt + \sigma(t)dB^1(t) \\ d(\ln \sigma(t)) = k(\theta - \ln(\sigma(t))dt + vdB^2(t) \end{cases}$$

Here, $S(t)$ is the asset price and $\sigma(t)$ is the instantaneous volatility at time t, and $\{B^1(t), B^2(t)\}$ are ordinary standard Brownian motions. Hull and White [7] have shown that, the price of European option at time t of exercise date T is the conditional expectation of the Black Scholes option pricing formula where the constant volatility from the original formula is replaced by the quadratic average over the period [t,T]:

(1.2)
$$\sigma_{t,T}^2 = \frac{1}{T-t} \int_t^T \sigma^2(u)d(u)$$

Such models are successful at capturing the symmetric smile and skewness of the implied volatility by imposing relations between the driving Brownian motions in (1.1) (symmetric smile is explained by independence while skewness can be explained by linear correlation).

Due to the temporal aggregation effect evident in [2], however, the smile effect deteriorates along with time-to-maturity since the temporal aggregation gradually erases the conditional heteroskedasticity; in the standard stochastic volatility setup, this particular decay is much faster than what is observed in market data. The phenomenon of slow decaying volatility smile is known as the volatility persistence (long-range dependence of volatility process). This phenomenon is particularly poignant for high frequency data, for which the conditional variance process displays near unit-root behavior.

Furthermore, we emphasize the existence of such phenomenon collaborated by large quantities of researches, pointing out that the conditional volatility of asset returns also displays long range dependence: [5], [6], [7], [8] have discussed extensively the evidence of such phenomenon in empirical data. Bayesian estimation in [9] of stochastic volatility models shows similar patterns of persistence.

Motivated by this inadequacy, long-memory process was deemed more appropriate enrichment for this purpose. Hence, fractional Brownian motion is a prime candidate among all long-memory process given its tractability and similarity with the ordinary Brownian motion: both the fractional Brownian motion and ordinary Brownian motion are self-similar with similar Gaussian structure. For discussions of estimation and evidence of the long-range dependence in conditional variance of asset returns, the reader is referred to [6] and section 3.1 of [10].

Now, we provide a formal definition of fractional Brownian motion (fBm). We adopt the definition as given in [11].

Definition 1.1 Let $H \in (0, 1)$ be a constant. A fractional Brownian motion $B^H(t)_{t \geq 0}$ with Hurst index H is a continuous and centered Gaussian process with covariance function

$$\mathbf{E}[\mathbf{B}^H(t)\mathbf{B}^H(s)] = \frac{1}{2}(t^{2H} + s^{2H} - |t - s|^{2H})$$

In particular, for $H = 1/2$, it reduces to the ordinary Brownian motion with $\mathbf{E}[B^H(t)B^H(s)] = \min(\mathbf{t}, \mathbf{s})$.

From equation (3) we have the following properties:

- $B^H(0) = 0$ and $E[B^H(t)] = 0, \forall t \geq 0$.

- $B^H(\cdot)$ has stationary increment: $B^H(t + s) - B^H(s)$ has the same distribution as $B^H(t)$ for any $s, t \geq 0$.

- $B^H(\cdot)$ is self-similar, meaning that $B^H(Tt)$ has the same distribution law as $(T)^H B^H(t)$.

- $B^H(\cdot)$ is a Gaussian process with the variance $\mathbf{E}[B^H(t)^2] = t^{2H}, \forall t \geq 0$.

- $B^H(\cdot)$ has continuous trajectories.

Fractional Brownian motions are divided into three very different categories: $H < 1/2, H = 1/2, H > 1/2$ .This is of particular importance because there is a deterministic difference between the case of $H < 1/2$ and $H > 1/2$, as we introduce the mathematical notion of long-range dependence.

Definition 1.2 (Long-range dependence) A stationary sequence $\{X_n\}_{(n \in N)}$ exhibits long-range dependence if the autocovariance function $\gamma(n) \doteq cov(X_k, X_{(k+n)})$ satisfies

$$\lim_{n \to \infty} \frac{\gamma(n)}{cn^{-\alpha}} = 1$$

This can be written as $(n) \sim |n|^{-\alpha}$,

In this case, for some constants c and $\alpha \in (0, 1)$, the dependence between X_k and X_{k+n} decays slowly as $n \to \infty$ and

$$\sum_{n=1}^{\infty} \gamma(n) = \infty$$

If we set $X_k = B^H(k) - B^H(k-1)$ and $X_{k+n} = B^H(k+n) - B^H(k+n-1)$ and apply equation (3), we have $\gamma_H(n) = \frac{1}{2}[(n+1)^{2H} + (n-1)^{2H} - 2n^{2H}]$ where $\gamma_H(n) = cov(B^H(k) - B^H(k-1), B^H(k+n) - B^H(k+n-1))$. In particular,

$$\lim_{n \to \infty} \frac{\gamma(n)}{H(2H-1)n^{2H-2}} = 1$$

Therefore, we conclude that: for $H > 1/2$, $\sum_{n=1}^{\infty} \gamma_H(n) = \infty$, and for $H > 1/2$, $\sum_{n=1}^{\infty} |\gamma_H(n)| < \infty$. Hence, only in the case of $H > 1/2$, fractional Brownian motions display long-memory dependence.

As pointed out in [12], large lags difference between $\gamma(\cdot)$ may be difficult to estimate in practice, so that models with long-range dependence are often formulated in terms of self-similarity. Self-similarity allows us to extrapolate across time scales and deduce long time behavior from short time behavior, which is more readily observed.

Because we are interested in capturing the long-memory phenomenon observed in financial markets, the rest of this chapter will only concern the case of $H > 1/2$.

1.2 Stochastic Integral Representation

In the original paper [2], Mandelbrot and van Ness represent the fBm in stochastic integral with respect to the ordinary Brownian motion:

(1.3)

$$B^H(t) = \frac{1}{\Gamma(H + \frac{1}{2})} \left(\int_{-\infty}^{0} \left[(t-s)^{H-\frac{1}{2}} - (-s)^{H-\frac{1}{2}} \right] dB(s) + \int_{0}^{t} (t-s)^{H-\frac{1}{2}} dB(s) \right)$$

where $\Gamma(\cdot)$ is the gamma function.

They have also included an alternative representation of the fBm which is the basis of the model in [13], [14]:

(1.4)
$$\widehat{B^H(t)} = \int_{0}^{t} \frac{(t-s)^{H-1/2}}{\Gamma(H+1/2)} dB(s)$$

This version of fBm is 'truncated' in the sense that the integration from negative infinity to zero in equation (1.3) is truncated. We will refer to the model (1.4) as the 'truncated fractional Brownian motion' in the rest of this chapter. As pointed out in [2], the representation (1.4) was first proposed by Paul Lévy to define the fBm by the Riemann-Liouville fractional integral, while the original integral in equation (1.3) is the Weyl fractional integral.

The definition (1.4) of fBm is in general not asymptotically covariance-stationary, even though it retains self-similarity. For further technical discussion and rigorous definition of the truncated fractional Brownian motion, we refer the reader to [13].

Given the differences, most of the analytical tools (such as Malliavin Calculus) developed for fBms might not be directly applicable to the truncated fBms. Yet, this truncated version of fBm is convenient to work in terms of simulation. These two types of fBms and their theoretical differences are covered in detail in [15].

2. Fractional Brownian Motions in Financial Models

We first look at several examples that utilize the fractional Brownian motions in the realm of financial modeling.

2.1 Asset Price Model

In the previous section, we mention that the motivation of fBms in finance models is to capture the long-range dependence in the volatility process. However, it is worth discussing the possibility of applying it to the asset process itself.

In practice, it is considered that an asset process driven by fBm will result in arbitrage. The idea is that, since for $H \neq 1/2$, $B^H(t)$ is not a semimartingale in continuous time, the asset process described by $B^H(t)$ violates the NFLVR (no free lunch with vanishing risk), a weaker version of arbitrage, and thus doesn't admit an equivalent martingale measure according to Theorem 7.2 of [16]. Such findings and construction of arbitrage can be found in Rogers [17].

In contrast, Cheridito [18] has given multiple classes of trading strategies that allow various level of arbitrages (NFLVR, arbitrage in the classical sense and strong arbitrage) under fBm driven assets, and shown that the arbitrages are all eliminated if the intra-transaction time is not zero, i.e., the classes of strategies become arbitrage-free. Such assumption is reasonable in practice, given the physical limit and non-zero transaction cost. For more information on arbitrage theory and its generalization, the reader is referred to [16], [17], [18].

2.2 Stochastic Volatility Model

As we have mentioned in the introductory section, the main motivation for fBm is to capture the volatility persistence, the stylized feature observed in empirical data. There are several prominent models involving a volatility process with fBm. Here, we just outline several of them.

2.2.1 Stochastic Volatility Model - Long-Memory Continuous Model

In [13], [14], Comte and Renault consider the following stochastic volatility model driven by fBm:

(2.1)
$$\begin{cases} \frac{dS(t)}{S(t)} &= rdt + \sigma(t)dB(t) \\ \sigma(t) &= \sigma_0 e^{x(t)} \\ dx(t) &= -kx(t)dt + v\widehat{dB^H}(t) \end{cases}$$

where the log-volatility term follows a fractional-OU process driven by the truncated fBm (1.4). The asset price process follows a geometric Brownian motion with volatility persistence.

Although Mandelbrot [2] deemed it as signifying the origin too heavily, the model (2.1) is easier and more robust than the ordinary fBms from the perspective of simulation. A simulation example is explored in Section 6.

2.2.2 Stochastic Volatility Model - Affine Fractional Stochastic Volatility Model

In [19], Comte et al. assume that the squared-volatility process follows

$$(2.2) \qquad \sigma^2(t) = \theta + X^{(\alpha)}(t)$$

where $X^{(\alpha)}(t)$ is defined by the fractional integral:

$$(2.3) \qquad X^{(\alpha)}(t) = \int_{\infty}^{t} \frac{(t-s)^{H-1/2}}{\Gamma(H+1/2)} X(s) ds$$

Similar treatment of the fractional integration is outlined in [13]. The affine structure in (2.2) is similar to the one originally studied by Duffie et al. [20].

The affine structure is adopted for the extra tractability, and thus better suited for practical option pricing and hedging than the original idea (2.1). In fact, Comte et al. [19] have shown that this model can better depict the difference between the short and long memory properties in the resulting option prices.

2.2.3 Stochastic Volatility Model - Martingale Expansion

Fukasawa [21] adopts and expands the asymptotic expansion technique first proposed by Yoshida [22] of European option prices around the Black-Scholes equation by means of perturbation technique and partial Malliavin calculus. It is shown that the logarithm of the asset process can be expressed as

$$lnS_t = Z_t = rt - \frac{1}{2} \int_0^t g(Y_s^n)^2 ds + \int_0^t g(Y_s^n) \left[\theta dW_s + \sqrt{1 - \theta^2} dW_s \right]$$

with

$$Y_s^n = y + \epsilon_n W_s^H, \qquad W_t^H = \int_0^t K_H(t, s) dW_s'$$

Here, r is the riskless rate, $\theta \in (-1, 1)$, $y \in R$ is a constant, (W, W') is a 2-dimensional standard (independent) Brownian motion, $\epsilon_n \to 0$ is a deterministic sequence for $n \to \infty$, and $g(\cdot)$ is the stochastic volatility process which is an adapted process for the minimal filtration.

Note that W_t^H is a fractional Brownian motion with Hurst index H, where $K_t^H(t, s)$ is the kernel of the stochastic integral representation over a finite interval of Brownian motion. According to [11], pertaining to our interest, for the case of $H > 1/2$, the kernel has the following expression:

$$K_H(t, s) = c_H s^{1/2-H} \int_s^t (u-s)^{H-3/2} u^{H-1/2} du$$

where

$$c_H = \left\{ \frac{H(2H-1)}{\beta(2-2H, H-1/2)} \right\}^{1/2}$$

Then, according to Corollary (2.6) of Fukasawa [21], the implied volatility can be expressed as

$$(2.4) \quad \sigma \left\{ 1 - \frac{\epsilon_n}{2} \rho_{13} d_2 \right\} + o\,(\epsilon_n) = aT^{H-1/2} \log(K/s) + \sigma + bT^{H+1/2} + o\,(\epsilon_n)$$

where d_2 is the typical argument in the $N(d_2)$ of the Black-Scholes formula, and

$$(2.5) \quad \rho_{13} = \frac{2\theta g'(y)c_H' T^H}{g(y)}, \quad \sigma = g(y), \quad a = \frac{\theta g'(y)c_H'}{\sigma}\epsilon_n, \quad b = -a\left(r - \frac{\sigma^2}{2}\right)$$

Equation (2.4) can be seen as an approximation for small $\epsilon_n \to 0$. Note however that, in order to check the accuracy of the approximation, we have no means but performing Monte Carlo simulation of fBm.

3. Simulation with Exact Methods

First, we look at methods that completely capture the covariance structure and true realization of the fractional Brownian motion (fBm) or fractional Gaussian noise (fGn). Any method described in this section has their starting point at the covariance matrix. The approximate scheme we see later is merely numerically close to the value of fBm (or fGn) or asymptotically coincides with it. The collection of algorithm in this section is not meant to be exhaustive. For more algorithm and discussion, see [23].

3.1 Hosking Method

The Hosking method utilizes the well-known conditional distribution of the multivariate Gaussian distribution on a recursive scheme to generate samples based on the explicit covariance structure. This method generates a general stationary Gaussian process with given covariance structure, not limited to generating fBms.

More specifically, this algorithm generates an fGn sequence $\{Z_k\}$ and fBm is recovered by accumulative sum. That is, the distribution of Z_{n+1} conditioned on the previous realization $Z_n, \ldots Z_1, Z_0$ can be explicitly computed.

Denote $\gamma(k)$ as the autocovariance function of the zero-mean process:

$$(3.1) \quad \gamma(k) \doteq \mathbf{E}(X_n X_{n+k})$$

where we assume for convenience that $\gamma(0) = 1$. For $n, k = 0, 1, 2 \ldots$, we have the following recursive relationship for the $(n+1) \times (n+1)$ autocovariance matrix $\Gamma(n) = \{\gamma(i-j)\}_{i,j=0,\ldots,n}$:

$$(3.2) \quad \begin{aligned} \Gamma(n+1) &= \begin{pmatrix} 1 & c(n)' \\ c(n) & \Gamma(n) \end{pmatrix} \\ &= \begin{pmatrix} \Gamma(n) & F(n)c(n) \\ c(n)'F(n) & 1 \end{pmatrix} \end{aligned}$$

where $c(n)$ is the $(n+1)$-column vector with elements $c(n)_k = \gamma(k+1), k = 0, \ldots, n$ and $F(n) = (\mathbf{1}(i = n - j))_{i,j=0,\ldots,n}$ is the $(n + 1) \times (n + 1)$ 'mirrored' identity matrix

$$F(n) = \begin{pmatrix} 0 & \cdots & 0 & 1 \\ 0 & \cdots & 1 & 0 \\ \vdots & \ddots & \vdots & \vdots \\ 1 & 0 & 0 & 0 \end{pmatrix}$$

Theorem 3.1 (Multivariate Gaussian distribution)

Any multivariate Gaussian random vector z can be partitioned into z_1 and z_2 with the mean vector and covariance matrix with the corresponding partition:

$$(3.3) \qquad \mu = \begin{bmatrix} \mu_1 \\ \mu_2 \end{bmatrix} \qquad \Sigma = \begin{bmatrix} \Sigma_{11} & \Sigma_{12} \\ \Sigma_{21} & \Sigma_{22} \end{bmatrix}$$

The distribution of z_1 conditioned on $z_2 = a$ is a multivariate normal $(z_1|z_2 = a) \sim N(\bar{\mu}, \bar{\Sigma})$ with

$$(3.4) \qquad \begin{aligned} \bar{\mu} &= \mu_1 + \Sigma_{12}\Sigma_{22}^{-1}(a - \mu) \\ \bar{\Sigma} &= \Sigma_{11} - \Sigma_{12}\Sigma_{22}^{-1}\Sigma_{21} \end{aligned}$$

If we substitute equation (3.2) into the partition in (3.4) with $\Sigma_{11} = 1, \mu = 0$, we have the following expression for the conditional distribution:

$$(3.5) \qquad \begin{aligned} \mu_{n+1} &= E\left(Z_{n+1}|Z_n, \cdots Z_0\right) = c(n)'\Gamma(n)^{-1}\begin{pmatrix} Z_n \\ \vdots \\ Z_1 \\ Z_0 \end{pmatrix} \\[2mm] \sigma_{n+1}^2 &= Var\left(Z_{n+1}|Z_n, \cdots Z_0\right) = 1 - c(n)'\Gamma(n)^{-1}c(n) \end{aligned}$$

With $Z_0 \sim N(0, 1)$, subsequently X_{n+1} for $n = 0, 1, \ldots$ can be generated.

Taking the inverse of $\Gamma(\cdot)$ at every step is computational expensive; the algorithm proposed by Hosking [24] computes the inverse $\Gamma(n + 1)^{-1}$ recursively. The next result is due to Dieker [23].

Proposition 3.1 (Hosking algorithm for simulating fGn) Define $d(n) = \Gamma(n)^{-1}c(n)$, and applying the blockwise method of inversion on equation (3.2):

$$(3.6) \qquad \Gamma(n + 1) = \frac{1}{\sigma_n^2}\begin{pmatrix} \sigma_n^2\Gamma(n)^{-1} + F(n)d(n)d(n)'F(n) & -F(n)d(n) \\ -d(n)'F(n) & 1 \end{pmatrix}$$

where σ_{n+1}^2 satisfies the recursion

$$(3.7) \qquad \sigma_{n+1}^2 = \sigma_n^2 - \frac{\gamma(n + 1) - \tau_{n-1})^2}{\sigma_n^2}$$

with $\tau_n = d(n)'F(n)c(n) = c(n)'F(n)d(n)$. Also, the recursion for $d(n+1) = \Gamma(n+1)^{-1}c(n+1)$ is obtained as

$$(3.8) \qquad d(n+1) = \begin{pmatrix} d(n) - \phi_n F(n)d(n) \\ \phi_n \end{pmatrix}$$

where

$$(3.9) \qquad \phi_n = \frac{\gamma(n+1) - \tau_n}{\sigma_n^2}$$

With $\mu_1 = \gamma(1)Z_0, \sigma_1^2 = 1 - \gamma(1)^2, \tau_0 = \gamma(1)^2, \mu_{n+1}, \sigma_{n+1}^2, \tau_{n+1}$ can be readily computed, and fractional Brownian motion is recovered by the cumulative summation.

This algorithm is also applicable to non-stationary processes (see [25] for details). Even though this algorithm is very simple and easy to understand and sample paths can be generated on-the-fly, the complexity of this algorithm is of $O(N^2)$ and computational (as well as memory) expense of this algorithm grows at a prohibitive speed.

3.2 Cholesky Method

Given that we are dealing with the covariance structure in matrix form, it is natural to go with the Cholesky decomposition: decomposing the covariance matrix into the product of a lower triangular matrix and its conjugate-transpose $\Gamma(n) = L(n)L(n)^*$. If the covariance matrix is proven to be positive-definite (the situation will be addressed in the next subsection), $L(n)$ will have real entries and $\Gamma(n) = L(n)L(n)'$.

Suppose that in matrix form the $(n+1) \times (n+1)$ product is given by

$$\begin{pmatrix} \gamma(0) & \gamma(1) & \gamma(2) & \cdots & \gamma(n) \\ \gamma(1) & \gamma(0) & \gamma(1) & \cdots & \gamma(n-1) \\ \gamma(2) & \gamma(1) & \gamma(0) & \cdots & \gamma(n-2) \\ \vdots & \vdots & \vdots & \ddots & \vdots \\ \gamma(n) & \gamma(n-1) & \gamma(n-2) & \cdots & \gamma(0) \end{pmatrix} = \begin{pmatrix} l_{00} & 0 & 0 & \cdots & 0 \\ l_{10} & l_{11} & 0 & \cdots & 0 \\ l_{20} & l_{21} & l_{22} & \ddots & \\ \vdots & \vdots & \vdots & \ddots & 0 \\ l_{n0} & l_{n1} & l_{n2} & \vdots & l_{nn} \end{pmatrix}$$
$$\times \begin{pmatrix} l_{00} & l_{10} & l_{20} & \cdots & l_{n0} \\ 0 & l_{11} & l_{21} & \cdots & l_{n1} \\ 0 & 0 & l_{22} & \cdots & l_{n2} \\ \vdots & \vdots & \vdots & \ddots & \vdots \\ 0 & 0 & 0 & 0 & l_{nn} \end{pmatrix}$$

It is easy to see that $l_{00}^2 = \gamma(0)$ and that $l_{10}l_{00} = \gamma(1)$ and $l_{10}^2 + l_{11}^2 = \gamma(0)$ on $i = 1$ (2nd row). For $i \geq 1$, the entries of the lower triangular matrix can be determined by

$$l_{i,0} = \frac{\gamma(i)}{l_{0,0}}$$

$$l_{i,j} = \frac{1}{l_{j,j}}\left(\gamma(i-j) - \sum_{k=0}^{j-1} l_{i,k}l_{j,k}\right), \qquad 0 < j \le n$$

$$l_{i,i}^2 = \gamma(0) - \sum_{k=0}^{i-1} l_{i,k}^2$$

Given independent, identically distributed (i.i.d.) standard normal random variables $(V_i)_{i=0,\ldots,n+1}$, the fGn sequence is generated by

$$Z_{n+1} = \sum_{k=0}^{n+1} l_{n+1,k}V_k$$

Or in matrix form, we have $Z(n) = L(n)V(n)$. If $\Gamma(n)$ is assumed to be positive-definite, the non-negativity of $l_{i,i}^2$ is guaranteed and $L(n)$ is guaranteed to be real. The covariance structure of the process is captured, since

(3.10) $Cov(Z(n)) = Cov(L(n)V(n)) = L(n)Cov(V(n))L(n)' = L(n)L(n)' = \Gamma(n)$

Even though the Cholesky method is easy to understand and implement, the computation time is $O(N^3)$, which renders this scheme extremely uneconomical in practice. To resolve this problem, we will proceed to another exact method. The idea is similar to retain the same relation as equation (3.10), but with a different decomposition.

3.3 Fast Fourier Transform Method

As we have seen from the last section, using the Cholesky decomposition seems to be the most straightforward idea to simulate Gaussian process with a given covariance structure; but, it also is the most rudimentary and thus slow. In order to improve upon the speed, the idea of utilizing the fast Fourier transform (FFT) was proposed by Davies and Harte [26] and further generalized by Dietrich and Newsam [27].

Similar to the idea before, this method tries to find a decomposition of the covariance matrix as $\Gamma = GG'$ and the sample is generated by $y = Gx$ for given standard normal random variable x. Then, on the given covariance structure, we have

$$Cov(y) = Cov(Gx) = GCov(x)G' = GG\prime = \Gamma$$

The idea is to 'embed' the original covariance matrix a circulant matrix in order to carry out the FFT. Before we outline the idea, we shall give out some detail of the linkage between Fourier transform and the circulant matrix.

Definition 3.1 (Circulant matrix) Circulant matrix is a special case of the Toeplitz matrix where each row vector is shifted to the right (the last element is shifted back to the beginning of the row). In matrix form, an n-by-n circulant matrix can be written as

$$C = \begin{pmatrix} c_0 & c_{n-1} & c_{n-2} & \cdots & c_1 \\ c_1 & c_0 & c_{n-1} & \cdots & c_2 \\ c_2 & c_1 & c_0 & \cdots & c_3 \\ \vdots & \vdots & \vdots & \ddots & \vdots \\ c_{n-1} & c_{n-2} & \cdots & c_1 & c_0 \end{pmatrix}$$

Remark: As one can see, the first row/column completely describes the whole matrix, and it can be put more succinctly in the following form:

$$c_{j,k} = c_{j-k(modn)}, \qquad \text{where } 0 \le j, k \le n - 1$$

Note that the indices range from 0 to $n - 1$ instead of the usual convention that ranges from 1 to n.

Definition 3.2 (Generating circulant matrix) We define an n-by-n generating circulant matrix by

$$G = \begin{pmatrix} 0 & 0 & 0 & \cdots & 0 & 1 \\ 1 & 0 & 0 & \cdots & 0 & 0 \\ 0 & 1 & 0 & \cdots & 0 & 0 \\ 0 & 0 & 1 & \cdots & 0 & 0 \\ \vdots & \vdots & \vdots & \ddots & \vdots & \vdots \\ 0 & 0 & 0 & \cdots & 1 & 0 \end{pmatrix}$$

By a simple calculation, we can see that the 'square' of the generating circulant matrix is given by

$$G^2 = \begin{pmatrix} 0 & 0 & 0 & \cdots & 1 & 0 \\ 0 & 0 & 0 & \cdots & 0 & 1 \\ 1 & 0 & 0 & \cdots & 0 & 0 \\ 0 & 1 & 0 & \cdots & 0 & 0 \\ \vdots & \vdots & \vdots & \ddots & \vdots & \vdots \\ 0 & 0 & \cdots & 1 & 0 & 0 \end{pmatrix}$$

From the point of view of row and column operation of the matrix, this can be seen as each row of the matrix being shifted one element forward, where the bumped element is replaced to the end of the row (it can also be thought of as the whole row is shifted down and the bumped row is placed back on top, but this is irrelevant to our interest). Arbitrary power can be deduced accordingly; this operation has a cycle of n iterations.

The generating circulant matrix is served as our building block. Looking back at our original circulant matrix, we have a corresponding polynomial

(3.11) $$p(x) = c_0 + c_1 x + c_2 x^2 + \cdots + c_{n-1} x^{n-1}$$

Then, the original circulant matrix C can be expressed as

$$
C = \begin{pmatrix}
c_0 & c_{n-1} & c_{n-2} & \cdots & c_2 & c_1 \\
c_1 & c_0 & c_{n-1} & \cdots & c_3 & c_2 \\
c_2 & c_1 & c_0 & \cdots & \vdots & c_3 \\
c_3 & c_2 & c_1 & \cdots & \vdots & \vdots \\
\vdots & \vdots & \vdots & \ddots & \vdots & c_{n-1} \\
c_{n-1} & c_{n-2} & c_{n-3} & \cdots & c_1 & c_0
\end{pmatrix}
$$

(3.12) $$C = p(G) = c_0 \mathbf{1} + c_1 G + c_2 G^2 + \cdots + c_{n-1} G^{n-1}$$

This can be verified by doing the row-operation of arbitrary power on G as shown above. It can be shown that each operation is one-element sub-diagonal compared to the previous power.

Definition 3.3 (Fourier matrix) The Fourier matrix is introduced as

$$
F = \begin{pmatrix}
1 & 1 & 1 & \cdots & 1 & 1 \\
1 & \xi & \xi^2 & \cdots & \xi^{n-2} & \xi^{n-1} \\
1 & \xi^2 & \xi^{2\times2} & \cdots & \vdots & \xi^{n-2} \\
1 & \xi^3 & \xi^{3\times2} & \cdots & \vdots & \vdots \\
\vdots & \vdots & \vdots & \ddots & \vdots & \xi^2 \\
1 & \xi^{n-2} & \xi^{n-2} & \cdots & \xi^2 & \xi
\end{pmatrix}
$$

Here, we define the n-th unity root as $\omega = e^{2\pi i \frac{1}{n}}$, and $\xi = \bar{\omega} = e^{-2\pi i \frac{1}{n}}$ is the conjugate of the unity root.

The Fourier matrix can be defined using the positive argument ω instead of ξ. Also, as we will see later, some definition includes the normalizing scalar $\frac{1}{\sqrt{n}}$ (or $\frac{1}{n}$). This is analogous to the continuous counterpart of the Fourier integral definition $F(f) = \int_{-\infty}^{+\infty} x(t) e^{-2\pi i f t} dt$ or $\int_{-\infty}^{+\infty} x(t) e^{2\pi i f t} dt$, as long as the duality is uphold by the opposite sign in the exponential power in the inverse Fourier transform. This duality will be restated in the diagonalization representation of the circulant matrix later.

Proposition 3.2 If $\frac{1}{\sqrt{n}}$ normalizes the Fourier matrix, then $\frac{1}{\sqrt{n}} F$ is a unitary matrix. It is symmetric (i.e., $F^T = F$), and the inverse of the Fourier matrix is given by

$$F^{-1} = \left(\frac{\sqrt{n}}{\sqrt{n}}F^{-1}\right) = \frac{1}{\sqrt{n}}\left(\frac{1}{\sqrt{n}}F\right)^{-1} = \frac{1}{\sqrt{n}}\left(\frac{1}{\sqrt{n}}\bar{F}^{T}\right) = \frac{1}{n}\bar{F} = \frac{1}{n}\begin{pmatrix} 1 & 1 & 1 & \cdots & 1 & 1 \\ 1 & \omega & \omega^2 & \cdots & \omega^{n-2} & \omega^{n-1} \\ 1 & \omega^2 & \omega^{2\times2} & \cdots & \vdots & \omega^{n-2} \\ 1 & \omega^3 & \omega^{3\times2} & \cdots & \vdots & \vdots \\ \vdots & \vdots & \vdots & \ddots & \vdots & \omega^2 \\ 1 & \omega^{n-1} & \omega^{n-2} & \cdots & \omega^2 & \omega \end{pmatrix}$$

Proposition 3.3 If we multiply the Fourier matrix with the generating circulant matrix, we have

$$FG = \begin{pmatrix} 1 & 1 & 1 & \cdots & 1 & 1 \\ 1 & \xi & \xi^2 & \cdots & \xi^{n-2} & \xi^{n-1} \\ 1 & \xi^2 & \xi^{2\times2} & \cdots & \vdots & \xi^{n-2} \\ 1 & \xi^3 & \xi^{3\times2} & \ddots & \vdots & \vdots \\ \vdots & \vdots & \vdots & \ddots & \vdots & \xi^2 \\ 1 & \xi^{n-1} & \xi^{n-2} & \cdots & \xi^2 & \xi \end{pmatrix}\begin{pmatrix} 0 & 0 & 0 & \cdots & 0 & 1 \\ 1 & 0 & 0 & \cdots & 0 & 0 \\ 0 & 1 & 0 & \cdots & 0 & 0 \\ 0 & 0 & 1 & \cdots & 0 & 0 \\ \vdots & \vdots & \vdots & \ddots & \vdots & \vdots \\ 0 & 0 & 0 & \cdots & 1 & 0 \end{pmatrix}$$

$$= \begin{pmatrix} 1 & 1 & \cdots & 1 & 1 & 1 \\ \xi & \xi^2 & \cdots & \xi^{n-2} & \xi^{n-1} & 1 \\ \xi^2 & \xi^{2\times2} & \cdots & \vdots & \xi^{n-2} & 1 \\ \xi^3 & \xi^{3\times2} & \ddots & \vdots & \vdots & 1 \\ \vdots & \vdots & \ddots & \vdots & \xi^2 & 1 \\ \xi^{n-1} & \xi^{n-2} & \cdots & \xi^2 & \xi & 1 \end{pmatrix}$$

This is the same as shifting (rotating) the first column to the back of the matrix, and is also equivalent to multiplying the first row with ξ^0, the 2nd row with ξ^1, etc. In matrix operation, it can be seen as

$$FG = \begin{pmatrix} \xi^0 & 0 & 0 & \cdots & 0 & 0 \\ 0 & \xi^1 & 0 & \cdots & 0 & 0 \\ 0 & 0 & \xi^2 & \cdots & \vdots & 0 \\ 0 & 0 & 0 & \ddots & \vdots & \vdots \\ \vdots & \vdots & \vdots & \ddots & \ddots & 0 \\ 0 & 0 & 0 & \cdots & 0 & \xi^{n-1} \end{pmatrix}\begin{pmatrix} 1 & 1 & \cdots & 1 & 1 & 1 \\ \xi & \xi^2 & \cdots & \xi^{n-2} & \xi^{n-1} & 1 \\ \xi^2 & \xi^{2\times2} & \cdots & \vdots & \xi^{n-2} & 1 \\ \xi^3 & \xi^{3\times2} & \ddots & \vdots & \vdots & 1 \\ \vdots & \vdots & \ddots & \vdots & \xi^2 & 1 \\ \xi^{n-1} & \xi^{n-2} & \cdots & \xi^2 & \xi & 1 \end{pmatrix} = \Lambda F$$

where Λ is the diagonal matrix with the k-th diagonal $\Lambda_k = \xi^k$, for $0 \le k \le n-1$. It follows that

(3.13) $$FGF^{-1} = \Lambda$$

That is, the Fourier matrix diagonalizes the generating circulant matrix with eigenvalues $\{\xi^k\}_{0 \le k \le n-1}$.

Theorem 3.2 The circulant matrix is decomposable by the Fourier matrix, i.e. $C = F^{-1} \Lambda F$ with eigenvalue matrix $\Lambda = \{p(\xi^k)\}_{k=0 \cdots n-1}$. Also, with equation (3.12), the diagonalization of C can be written as

$$
\begin{aligned}
FCF^{-1} &= F \left(c_0 \mathbf{1} + c_1 G + c_2 G^2 + \cdots + c_{n-1} G^{n-1} \right) F^{-1} \\
&= c_0 \mathbf{1} + c_1 \left(FGF^{-1} \right) + c_2 \left(FGF^{-1} \right)^2 + \cdots + c_{n-1} \left(FGF^{-1} \right)^{n-1} \\
&= \begin{pmatrix}
p(1) & 0 & 0 & \cdots & 0 & 0 \\
0 & p(\xi) & 0 & \cdots & 0 & 0 \\
0 & 0 & p(\xi^2) & \cdots & \vdots & 0 \\
0 & 0 & 0 & \ddots & \vdots & \vdots \\
\vdots & \vdots & \vdots & \ddots & \vdots & 0 \\
0 & 0 & 0 & \cdots & 0 & p(\xi^{n-1})
\end{pmatrix}
\end{aligned}
$$

Note that $(FGF^{-1})^2 = FGF^{-1} FGF^{-1} = FGGF^{-1} = FG^2 F^{-1}$. The other powers can be deduced iteratively.

This theorem gives us the fundamental theoretical framework to build up the FFT exact simulation of fBms. The basic idea of the simulation is to embed the covariance matrix into a bigger circulant matrix to carry out the discrete Fourier transform as outlined above (with technique of FFT). Such technique is called Circulant Embedding Method (CEM), and is outlined in Dietrich and Newsam [27] and Perrin et al. [28].

Suppose that we need sample size of N (N should be a power of 2, i.e. $N = 2^g$ for some $g \in N$ for the sake of convenience when facilitating FFT). Generate the N-by-N covariance matrix Γ with entries $\Gamma_{j,k} = \gamma(|j - k|)$, where γ is the covariance function given in the definition of fractional Gaussian noise (fGn), by

$$
\Gamma = \begin{pmatrix}
\gamma(0) & \gamma(1) & \cdots & \gamma(N-1) \\
\gamma(1) & \gamma(0) & \cdots & \gamma(N-2) \\
\vdots & \vdots & \ddots & \vdots \\
\gamma(N-1) & \gamma(N-2) & \cdots & \gamma(0)
\end{pmatrix}
$$

The technique to simulate fGn with FFT is called the Circulant Embedding Method (CEM), generalized by Davies and Harte [26], and consists of embedding this covariance matrix into a bigger M-by-M (with $M = 2N$) circulant covariance matrix C such as

$$C =$$

$$
\begin{pmatrix}
\gamma(0) & \gamma(1) & \cdots & \gamma(N-1) & \gamma(0) & \gamma(N-1) & \gamma(N-2) & \cdots & \gamma(2) & \gamma(1) \\
\gamma(1) & \gamma(0) & \cdots & \gamma(N-2) & \gamma(N-1) & 0 & \gamma(N-1) & \cdots & \gamma(3) & \gamma(2) \\
\vdots & \vdots & \ddots & \vdots & \vdots & \vdots & \vdots & \ddots & \vdots & \vdots \\
\gamma(N-1) & \gamma(N-2) & \cdots & \gamma(0) & \gamma(1) & \gamma(2) & \gamma(3) & \cdots & \gamma(N-1) & 0 \\
0 & \gamma(N-1) & \cdots & \gamma(1) & \gamma(0) & \gamma(1) & \gamma(2) & \cdots & \gamma(N-2) & \gamma(N-1) \\
\gamma(N-1) & 0 & \cdots & \gamma(2) & \gamma(1) & \gamma(0) & \gamma(1) & \cdots & \gamma(N-3) & \gamma(N-2) \\
\vdots & \vdots & \ddots & \vdots & \vdots & \vdots & \vdots & \ddots & \vdots & \vdots \\
\gamma(1) & \gamma(2) & \cdots & 0 & \gamma(N-1) & \gamma(N-2) & \gamma(N-3) & \cdots & \gamma(1) & \gamma(0)
\end{pmatrix}
$$

where the covariance matrix is embedded on the top left hand corner. It is sufficient to point out that

$$
C_{0,k} = \begin{cases} \gamma(k) & k = 0, \cdots, N-1 \\ \gamma(2N-k) & k = N+2, \cdots, 2N-1 \end{cases}
$$

Remark: As Perrin et al. [28] have pointed out, the size M can be $M \geq 2(N-1)$, and the case $M = 2(N-1)$ is minimal embedding. For any other choice of M, the choice of $C_{0,N}, \ldots, C_{0,M-N+1}$ is arbitrary and can be conveniently chosen as long as the symmetry of the matrix is upheld; more zeros can be padded if M is bigger to make C circulant. For the rest of the chapter, we will concern ourselves with the case $M = 2N$.

From Theorem 3.2, we know that, given any circulant matrix, it can be decomposed as $C = Q\Lambda Q^*$, where

$$(3.14) \qquad (Q)_{j,k} = \frac{1}{\sqrt{2N}} \exp\left(-2\pi i \frac{jk}{2N}\right) \qquad , \text{for } j, k = 0, \cdots, 2N-1$$

The matrix Λ is the diagonal matrix with eigenvalues

$$(3.15) \qquad \lambda_k = \sum_{j=0}^{2N-1} c_{0,j} \exp\left(2\pi i \frac{jk}{2N}\right) \qquad , \text{for } j, k = 0, \ldots, 2N-1$$

This differs slightly from the previous definition, but similar to the continuous counterpart; the sign of the exponential power in the Fourier transform is just conventional difference. The approach is identical as long as the duality is maintained. That is, if written in the form of $C = Q\Lambda Q^*$, the sign of the exponential power of the component in Q and Λ should be opposite. In the case of the previous theorem where $C = F^{-1}\Lambda F$, it is easy to check that F^{-1} and Λ (ξ) indeed have the opposite sign in the exponential power.

It should be noted that C is not guaranteed to be positive-definite. Davies and Harte [26] suggest setting zero every negative value that may appear in Λ. In Perrin et al. [30], they prove that the circulant covariance matrix for fGn is always non-negative definite, so the approach is feasible without any modification. The reader is referred to Dietrich and Newsam [27] and Wood and Chan [29] for more detail on dealing with this issue.

Assuming that C is positive definite and symmetric, the eigenvalues are positive and real. The 'square root' of C is readily formed, $S = Q\Lambda^{1/2}Q^*$, where $\Lambda^{1/2}$ is the diagonal matrix with eigenvalues $1, \sqrt{\lambda_1}, \cdots, \sqrt{\lambda_{2N-1}}$. It is easy to check that $SS^* = SS' = C$. So, S has the desired properties we look for.

Theorem 3.3 (Simulation of fGn with FFT) The simulation of the sample path of fGn, we are going to simulate $y = SV$, consists of the following steps:

1. Compute the eigenvalues $\{\lambda_k\}_{k=0,\ldots,2N-1}$ from equation (3.15) by means of FFT. This will reduce the computational time from $O(N^2)$ to $O(N \log N)$.

2. Calculate $W = Q^*V$.

   - Generate two standard normal random variables $W_0 = V_0^{(1)}$ and $W_N = V_N^{(1)}$

   - For $1 \le j < N$, generate two standard normal random variables $V_j^{(1)}$ and $V_j^{(2)}$ and let

     - $W_j = \frac{1}{\sqrt{2}}\left(V_j^{(1)} + iV_j^{(2)}\right)$
     - $W_{2N-j} = \frac{1}{\sqrt{2}}\left(V_j^{(1)} - iV_j^{(2)}\right)$

3. Compute $Z = Q\Lambda^{1/2}W$. This can be seen as another Fourier transform of the vector $\Lambda^{1/2}W$:

   - $Z_k = \frac{1}{\sqrt{2N}} \sum_{j=0}^{2N-1} \sqrt{\lambda_j} W_j \exp\left(-2\pi i \frac{jk}{2N}\right)$

   - It is identical to carry out FFT on the following sequence:

   $$w_j = \begin{cases} \sqrt{\frac{\lambda_0}{2N}} V_0^{(1)} & j = 0 \\ \sqrt{\frac{\lambda_j}{4N}}\left(V_j^{(1)} + iV_j^{(2)}\right) & j = 1, \cdots, N-1 \\ \sqrt{\frac{\lambda_N}{2N}} V_N^{(1)} & j = N \\ \sqrt{\frac{\lambda_j}{4N}}\left(V_{2N-j}^{(1)} - iV_{2N-j}^{(2)}\right) & j = N+1, \cdots, 2N-1 \end{cases}$$

   - Due to the symmetric nature of the sequence, the Fourier sum of $\{w_j\} = \{Z_k\}_{k=0}^{2N-1}$ is real. The first N samples have the desired covariance structure. But, since the 2nd half of samples $(N \cdots 2N - 1)$ are not independent of the first N samples, this sample cannot be used.

4. Recover fBm from the recursive relationship:

- $B^H(0) = 0;$ $\qquad B^H(i) = B^H(i-1) + Z_{i-1},$ $\qquad$ for $1 \le i \le N$

4. Approximate Methods

As we have seen in the previous section, the exact methods all take on the covariance structure matrix as starting point and try to reproduce the covariance structure by different decomposition. That can be time and resource consuming, so rather it is preferable to have approximation of the fractional Brownian motion that permits robust simulation.

In this section, we will start with the Mandelbrot representation due to historical reason and move onto several other methods that provide us with better understanding of the process and increasing robustness.

4.1 Mandelbrot Representation

Recalling from Section 1.2, fractional Brownian motion permits a stochastic integral representation. To approximate equation (1.3), it is natural to truncate the lower limit from negative infinity to some point, say at $-b$:

$$(4.1) \quad \widetilde{B^H}(n) = C_H \sum_{k=-b}^{0} \left[(n-k)^{H-1/2} - (-k)^{H-1/2} \right] B_1(k) + \sum_{k=0}^{n} (n-k)^{H-1/2} B_2(k)$$

Note that the C_H is not the same constant term as in equation (1.3), because one has to re-calculate the normalizing factor due to the truncation. As pointed out in [23], the recommended choice of b is $N^{3/2}$. Even though this is a straightforward way to generate fractional Brownian motion, it is rather inefficient. It is included in this section for the sake of completeness.

4.1.1 Euler Hypergeometric Integral

fBm permits the stochastic integral form involving the Euler hypergeometric integral:

$$B^H(t) = \int_0^t K_H(t,s)dB(s)$$

where $B(s)$ is the standard Brownian motion and

$$K_H(t,s) = \frac{(t-s)^{H-1/2}}{\Gamma(H+1/2)} F_{2,1}\left(H - \frac{1}{2}; \frac{1}{2} - H; H + \frac{1}{2}; 1 - \frac{t}{s} \right)$$

is the hypergeometric function, which can be readily computed by most mathematical packages. By discretizing (28), at each time index t_j, we have

$$B^H(t_j) = \frac{n}{T} \sum_{i=0}^{j-1} \left(\int_{t_i}^{t_{i+1}} K_H(t_j, s) ds \right) \delta B_i$$

where $\delta B_i = \sqrt{\frac{T}{n}} \Delta B_i$ and ΔB_i is drawn according to the standard normal distribution. This means that $\{\delta B_i\}_{i=1\cdots n}$ is the increments of a scaled Brownian motion on $[0, T]$ with quadratic variation T. The inner integral can be computed by the Gaussian quadrature efficiently.

4.2 Construction by Correlated Random Walk

This particular algorithm proposed in [30] of constructing fBm relies on the process of correlated random walks and summation over generated paths. This is similar to the generation of ordinary Brownian method through summation of the sample paths of normal random walk, which is related to the central limit theorem.

Definition 4.1 (Correlated Random Walk) For any $p \in [0, 1]$, denote X_p^n as the correlated random walk with persistence index p. It consists of a jump on each time-step with jump size of either $+1$ or -1 such that:

- $X_0^p = 0, P(X_1^p = -1) = 1/2, P(X_1^p = +1) = 1/2$

- $\forall n \geq 1, \epsilon_n^p \equiv X_n^p - X_{n-1}^p$ which equals either $+1$ or -1

- $\forall n \geq 1, P(\epsilon_{n+1}^p = \epsilon_n^p | \sigma(X_k^p, 0 \leq k \leq n)) = p$

Theorem 4.1 For any $m \geq 1, n \geq 0$, we have

$$(4.2) \qquad \mathbf{E}\left[\epsilon_m^p \epsilon_{m+n}^p\right] = (2p - 1)^n$$

In order to add additional randomness into the correlated random walks, we replace the constant persistence index p with a random variable μ, and we denote the resulting correlated random walk as X_n^μ. Or, to put it more formally, denote by P^p the law of X^p for a given persistence index p. Now, consider a probability measure μ on $[0, 1]$, which we call the corresponding probability law P^μ, the annealed law of the correlated walk associated to μ. Note that $P^\mu \triangleq \int_0^1 P^p d\mu(p)$

Proposition 4.1 For all $m \geq 1, n \geq 0$, we have

$$(4.3) \qquad \mathbf{E}\left[\epsilon_m^\mu \epsilon_{m+n}^\mu\right] = \int_0^1 (2p - 1)^n \, d\mu(p)$$

The next result is due to Enriquez [30]. The proof is based on Lemma 5.1 of Taqqu [31].

Theorem 4.2 For $1/2 < H < 1$, denote by μ^H the probability on $[1/2, 1]$ with density $(1 - H) 2^{3-2H} (1 - p)^{1-2H}$. Let $\left(X^{\mu^H, i}\right)_{i \geq 1}$ be a sequence of independent processes with probability law P^{μ^H}. Then,

$$(4.4) \qquad \mathcal{L}^D \lim_{N \to \infty} \mathcal{L} \lim_{N \to \infty} c_H \frac{X_{[Nt]}^{\mu^H, 1} + \cdots + X_{[Nt]}^{\mu^H, M}}{N^H \sqrt{M}} = B_H(t)$$

where $c_H = \sqrt{\frac{H(2H-1)}{\Gamma(3-2H)}}$, $\mathcal{L}$ stands for the convergence in law, and $\mathcal{L}^D$ means the convergence in the sense of weak convergence in the Skorohod topology on $D[0,1]$, the space of cadlag functions on $[0,1]$. Here, $[\cdot]$ is the floor function and rounds the argument to the closest integer, M is the number of trajectories of correlated random walks and N is number of time steps.

Remark: For $H = 1/2$, there is a similar expression (see [30]). The order of limit in equation (4.4) is important, because if reversed, the sum would result in 0. Theorem 4.2 is mostly for theoretical construction.

In [30], the above theorem is further simplified from double summations into a single summation by applying Berry-Essen bound: As long as $M(N)$ is of order $O\left(N^{2-2H}\right)$,

$$(4.5) \qquad \mathcal{L}^D \lim_{N \to \infty} c_H \frac{X_{[Nt]}^{\mu^H,1} + \cdots + X_{[Nt]}^{\mu^H,M(N)}}{N^H \sqrt{M(N)}} = B_H(t)$$

In practice, any probability measure with moment equivalent to $\frac{1}{n^{2-2H}L(n)}$, where L is a slowly varying function, will be used. This could be shown by Karamata's theorem, for which further elaboration is found in [30]. In [30], three families of equivalent measures are provided, and specifically the 2nd family of the measures $\left(\mu'_{H,k}\right)_{k>0}$ is most appropriate for simulation purpose:

For $H > 1/2$, $\mu'_{H,k}$ has the density of $1 - \frac{(1-U^{1/k})^{\frac{1}{2-2H}}}{2}$ with the corresponding normalizing factor $c'_{H,k} = \frac{c_H}{\sqrt{k}}$. The error given by the Berry-Essen bound for $H > 1/2$ is given by

$$(4.6) \qquad 0.65 \times D_H N^{1-H} / \sqrt{kM}$$

where $D_H = \sqrt{\frac{6(2H-1)}{(H+1)(2H+1)}}$. Here, k serves as a control variable of order $k(N) = o(N)$, and the error term contains $\frac{1}{\sqrt{k}}$ which can be seen as a way to restrict error with the price of distortion of the covariance relation in X_N, though it is advisable to keep $k \le 1$. For more discussion on the choice of k, we refer to Section 4 of [30].

Theorem 4.3 (Simulation of fBm with correlated random walk) Simulating fBm with correlated random walk for the case of $H > 1/2$ consists of the following steps:

1. Calculate $M(N)$ by the tolerable error level from equation (4.6). Calculate $\lfloor NT \rfloor$, where $\lfloor \cdot \rfloor$ is the floor function, and create time-index t_i : $\{1, 2, \cdots, \lfloor NT \rfloor\}$.

2. Simulate M independent copies of $\left\{\mu_{H,k}^j\right\}_{j=1\cdots M} = 1 - \frac{(1-U^{1/k})^{\frac{1}{2-2H}}}{2}$ for M trajectories.

3. Simulate M copies of correlated random walks:

- If $t_i = 1$, $\epsilon_1^j = 2 * Bernoulli(\frac{1}{2}) - 1$, $\quad X_1^j = \epsilon_1^j$
- If $t_i > 1$, $\epsilon_{t_j}^j = \epsilon_{t_{j-1}}^j * \left(2 * Bernoulli\left(\mu_{H,k}^j\right) - 1\right)$, $\quad X_{t_j}^j = X_{t_{j-1}}^j + \epsilon_{t_j}^j$

4. At each t_j, calculate

- $$B_{t_j}^H = c_H' \frac{X_{\left[Nt_j\right]}^{\mu^{H,1}} + \cdots + X_{\left[Nt_j\right]}^{\mu^{H,M}}}{N^H \sqrt{M}}$$

Remark: For any given time-horizon T, it is easier to simulate the path of $B^H(1)$ with given resolution N and scale it to arrive at $B^H(T) = T^H B^H(1)$.

This algorithm is interesting from a theoretical point of view, since it gives us a construction of fractional Brownian motion reflecting its ordinary Brownian motion counterpart with the help of central limit theorem. But, in practice, it might not be fast enough for simulation purpose that requires large number of simulated paths.

4.3 Conditional Random Midpoint Displacement

This algorithm is put forth by Norros et al. [32] and uses the similar approach to compute the conditional distribution of the fractional Gaussian noises as we have seen in the Hosking algorithm in Section 3.1. The difference is that this algorithm does not completely capture the covariance of all the sample points, instead it chooses certain number of points of the generated samples and uses a different ordering to do conditioning on (recall that, in the Hosking method, the conditioning is done by chronological order).

Again we are interested in the stationary fractional Gaussian noises and will back out the fractional Brownian motion on the time interval [0,1], which can be scaled back according to self-similarity relationship. We will first outline the idea of bisection method, which will be expanded into the conditional mid-point replacement scheme later on.

4.3.1 Bisection Scheme and Basic Algorithm

In this section we adopt the notation in Norros et al. [32]: $Z(t)$ is the fractional Brownian motion, and $X_{i,j}$ is the fractional Gaussian noise of a certain interval j in a given level i.

The idea is to simulate $Z(t)$ on the interval $[0, 1]$. First, given $Z(0) = 0$ and $Z(1)$ with the standard normal distribution of $N(0, 1)$, we compute the conditional distribution of $\{Z(1/2)|Z(0), Z(1)\} \sim N(\frac{1}{2}Z(1), 2^{-2H} - \frac{1}{4})$. The bisection involves the indices i and j, where i indicates the 'level' and j for the 'position. Let

$$X_{i,j} = Z\left(j \cdot 2^{-i}\right) - Z\left((j-1) \cdot 2^{-i}\right), \qquad \text{for } i = 0, 1, 2, \ldots, j = 1, \cdots, 2^i$$

It is easy to see that, at any given level i, the interval $[0, 1]$ is divided into 2^i sub-intervals.

If we denote $(i-1)$th level as the 'mother-level', it will be divided twice finer in the next level. So given any interval on the mother level, it is easy to observe the relationship

$$(4.7) \qquad X_{i,2j-1} + X_{i,2j} = X_{i-1,j}$$

Because of equation (4.7), it is enough to just generate $X_{i,j}$ for odd number j. So, let us proceed from left to right, assuming that the sample points $X_{i,1}, \cdots X_{i,2k}$ have already been generated $k \in \{0, 1, \ldots, 2^{i-1} - 1\}$. For the point $X_{i,2k+1}$, we have
(4.8)
$$X_{i,2k+1} = \mathbf{e}(i,k)\left[X_{i,\max(2k-m+1,1)}, \cdots, X_{i,2k}, X_{i-1,k+1}, \cdots, X_{i-1,\min(k+n,2^{i-1})}\right]' + \sqrt{v(i,k)}U_{i,k}$$

where $U_{i,k}$ are i.i.d. standard Gaussian random variables $i = 0, 1, \cdots$; $k = 0, 1, \cdots, 2^{i-1} - 1$. Equation (4.8) can be rewritten as
(4.9)
$$X_{i,2k+1} = \mathbf{e}(i,k)\left[X_{i,(2k-m+1,1)}, \cdots, X_{i,2k}, X_{i-1,k+1}, \cdots, X_{i-1,\min(k+n,2^{i-1})}\right]' + \sqrt{v(i,k)}U_{i,k}$$
$$= \mathbf{E}\left[X_{i,2k+1}|X_{i,\max(2k-m+1,1)}, \cdots, X_{i,2k}, X_{i-1,k+1}, \cdots, X_{i-1,\min(k+n,2^{i-1})}\right]$$

(4.10)
$$v(i,k) = Var\left[X_{i,2k+1}|X_{i,\max(2k-m+1,1)}, \cdots, X_{i,2k}, X_{i-1,k+1}, \cdots, X_{i-1,\min(k+n,2^{i-1})}\right]$$

As mentioned before, this scheme conditions on a fixed number of past samples instead of the whole past, where the two integers ($m \geq 0, n \geq 1$) indicate the number of the intervals the expectation and variance are conditioned on, and is called RMD(m,n). Looking at (4.9) and (4.10) "$X_{i,(2k-m+1,1)}, \cdots, X_{i,2k}$" indicates that there are at most m neighboring increments to the left of the interval in question $X_{i,2k+1}$, and "$X_{i-1,k+1}, \cdots, X_{i-1,\min(k+n,2^{i-1})}$" indicates that there are at most n neighboring increments to the right of the interval.

Denote by Γ_{ik} the covariance matrix with $X_{i,2k+1}$ as the first entry, and $X_{i,\max(2k-m+1,1)}, \cdots, X_{i,2k}, X_{i-1,k+1}, \cdots, X_{i-1,\min(k+n,2^{i-1})}$ as the rest of the entries. Then, we have

$$\Gamma_{ik} = Cov\left[X_{i,2k+1}, X_{i,\max(2k-m+1,1)}, \cdots, X_{i,2k}, X_{i-1,k+1}, \cdots, X_{i-1,\min(k+n,2^{i-1})}\right]$$

where

$$Cov([x_1, x_2, \cdots, x_n]) = \begin{pmatrix} Cov(x_1, x_1) & Cov(x_1, x_2) & \cdots & Cov(x_1, x_n) \\ Cov(x_2, x_1) & Cov(x_2, x_2) & \cdots & Cov(x_2, x_n) \\ \vdots & \vdots & \ddots & \vdots \\ Cov(x_n, x_1) & Cov(x_n, x_2) & \cdots & Cov(x_n, x_n) \end{pmatrix}$$

Similar to (3.3)-(3.5), we can partition Γ_{ik} as

$$\Gamma_{ik} = \begin{pmatrix} Var(X_{i,2k+1}) & \Gamma_{ik}^{(1,2)} \\ \Gamma_{ik}^{(2,1)} & \Gamma_{ik}^{(2,2)} \end{pmatrix}$$

Note that $\Gamma_{ik}^{(1,2)} = (\Gamma_{ik}(2,1))'$. Hence, we have

(4.11)
$$\mathbf{e}(i,k) = \Gamma_{ik}^{(1,2)} \left(\Gamma_{ik}^{(2,2)}\right)^{-1}$$

(4.12)
$$v(i,k) = \frac{|\Gamma_{ik}|}{\left|\Gamma_{ik}^{(2,2)}\right|}$$

By the stationarity of the increment of Z and by self-similarity, $\mathbf{e}(i,k)$ is independent of i and k when $2k \geq m$ and $k \leq 2^{i-1} - n$ (meaning that the sequence is not truncated by $\max(\cdot)$ and $\min(\cdot)$, and it only depends on i only when $2^i < m + 2n$.

4.3.2 On-the-Fly RMD(m,n) Generation

Norros et al. [32] further propose that, instead of having the previous level $(i-1)$ completely generated first, partition and conditioning can be done 'on-the-fly', meaning that we can have multiple unfinished levels going at the same time. Unlike the previous RMD(m,n) scheme, the level here is defined differently.

First define the 'resolution' by δ, as the smallest interval that we will be dealing with in this scheme. Note that this is different from the previous sub-section where at the i-th level $\delta = 2^{-i}$, which can be bisected finer indefinitely. In the on-the-fly RMD scheme, the minimum interval length is defined as δ.

At each level i, the interval $\left[0, 2^i\delta\right]$ is split finely into interval with length of δ and $X_{i,k}$ samples are generated on each point until all points within the interval are filled. Then, the trace is expanded to the next level $i+1$, the interval $\left[0, 2^{i+1}\right]$, and this procedure can be considered as 'enlargement'. So, instead of having a pre-determined time-horizon and zooming in with twice-finer resolution in the original RMD scheme, the new RMD scheme has a pre-determined resolution and expand twice-fold the horizon at each level.

Within the same interval $\left[0, 2^i\right]$, the following rules are applied for the inner intervals:

1. Must have $n-1$ (or all) right-neighboring intervals (of which, have the same length as the mother-interval).

2. Also, there should be m (or all) left-neighboring intervals (of which, have the same length as the interval being considered).

3. If there are intervals that satisfy both the conditions above, choose the one as left as possible.

4. When all intervals are filled out, expand to the next level by 'enlargement'.

Here we use

$$Y_{i,j} = Z\left(j \cdot 2^i\delta\right) - Z\left((j-1) \cdot 2^i\delta\right), \quad i = 0, 1, \cdots, \quad j = 1, 2, \cdots$$

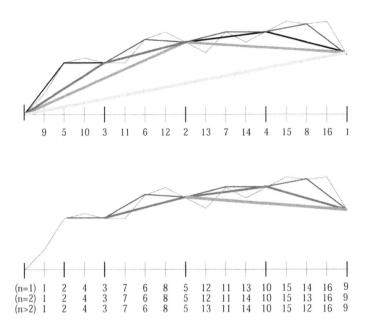

Figure 1. The order of splits by the ordinary RMD(m,n) scheme (top) and the on-the-fly RMD(m,n) scheme (bottom). Note that, for the on-the-fly scheme, the order changes according to the choice of n.

instead of $X_{i,j}$ to avoid confusion with the original RMD scheme notation.

Similar to the ordinary RMD, we have the following equations from the conditional distribution of multivariate Gaussian processes: The enlargement stage is defined by

$$\begin{aligned} Y_{i,1} &= \mathbf{E}\left[Y_{i,1}|Y_{i-1,1}\right] + \sqrt{Var\left[Y_{i,1}|Y_{i-1,1}\right]}U_{i,1} \\ &= \mathbf{e}\,(i,1)\,Y_{i-1,1} + \sqrt{v\,(i,1)}U_{i,1} \end{aligned}$$

Inner interval points are constructed similar to the ordinary RMD scheme (the right-neighboring intervals are of level $i + 1$ instead of $i - 1$) as

$$Y_{i,2k+1} = \mathbf{e}\,(i,k)\left[Y_{i,\max(2k-m+1,1)}, \cdots, Y_{i,2k}, Y_{i+1,k+1}, \cdots, Y_{i+1,\min(k+n,N_{i+1})}\right]' + \sqrt{v\,(i,k)}U_{i,k}$$

where N_{i+1} is the last generated increment on the $(i + 1)$-th level. The order of splitting for the on-the-fly scheme is given in Figure 1, where its ordinary RMD counterpart's splitting order is also given.

Norros et al. [32] have done an extensive comparison between on-the-fly RMD schemes in terms of accuracy and robustness compared to the FFT and aggregate methods. On-the-fly RMD and FFT are significantly faster than the aggregate method, and on-the-fly RMD can generate samples with no fixed time-horizon, while for FFT the whole trace has to be generated before it can be used. So, RMD seems superior in terms of flexibility.

4.4 Spectral Method

In this subsection, we will look into the spectral method of approximating the fractional Gaussian noises, which has the origin from spectral analysis in physics: A time-domain can be transformed into a frequency-domain without loss of information through Fourier transform. With the typical Fourier-time series, the original input is deterministic and transformed into the spectral density that represents the magnitude of different frequencies in the frequency domain. It is possible to extend this approach to analyzing stochastic processes. Though it is impossible to study all realization, it is possible to analyze in a probabilistic/distribution sense by observing the expected frequency information contained in the autocovariance function.

Spectral density is computed for frequencies, $-\pi \leq \lambda \leq \pi$, as

$$f(\lambda) = \sum_{j=-\infty}^{\infty} \gamma(j) \exp(ij\lambda)$$

The $\gamma(\cdot)$ here is the autocovariance function, which can be recovered by the inverse formula

$$\gamma(\cdot) = \frac{1}{2\pi} \int_{-\pi}^{\pi} f(\lambda) \exp(-ij\lambda)$$

The spectral density of the fGn can be approximated according to [23] and [33] as

$$f(\lambda) = 2 \sin(\pi H) \Gamma(2H + 1)(1 - \cos \lambda) \left[|\lambda|^{-2H-1} + B(\lambda, H) \right]$$

where

$$B(\lambda, H) = \sum_{j=1}^{\infty} \left\{ (2\pi j + \lambda)^{-2H-1} + (2\pi j - \lambda)^{-2H-1} \right\}$$

Note that the domain is only $-\pi \leq \lambda \leq \pi$, since any frequency higher would only correspond to amplitude between our desired sample points.

The problem with the above expression is that there is no known form for $B(\lambda, H)$, Paxson [35] proposes the following scheme for $B(\lambda, H)$:

$$B(\lambda, H) \cong \tilde{B}_3(\lambda, H) = a_1^d + b_1^d + a_2^d + b_2^d + a_3^d + b_3^d + \frac{a_3^{d'} + b_3^{d'} + a_4^{d'} + b_4^{d'}}{8H\pi}$$

where

$$d = -2H - 1, \quad d' = -2H, \quad a_k = 2k\pi + \lambda, \quad b_k = 2k\pi - \lambda$$

Moreover, with the help of the Whittle estimator, Paxson [33] shows that

$$\tilde{B}_3(\lambda, H)'' = [1.0002 - 0.000134\lambda] \left(\tilde{B}_3(\lambda, H) - 2^{-7.65H - 7.4} \right)$$

gives a very robust and unbiased approximation for the $B(\lambda, H)$. See Appendix A of [33] for a detailed comparison and justification of this approximation.

With the approximation scheme for the spectral density at hand, we can now look at the spectral analysis of a stationary discrete-time Gaussian process (fractional Gaussian noise; fGn) $X = \{X_n : n = 0, \cdots, N - 1\}$, which can be represented in terms of the spectral density $f(\lambda)$ as

$$X_n = \int_0^\pi \sqrt{\frac{f(\lambda)}{\pi}} \cos(n\lambda) \, dB_1(\lambda) - \int_0^\pi \sqrt{\frac{f(\lambda)}{\pi}} \sin(n\lambda) \, dB_2(\lambda)$$

where B_1 and B_2 are independent standard Brownian motions and the equality is understood in terms of distribution. Define $\xi_n(\lambda) = \sqrt{\frac{f(\lambda)}{\pi}} \cos(n\lambda)$ and fix some integer l. After setting $t_k = \frac{\pi k}{l}$ for $k = 0, \cdots, l - 1$, we can approximate it by a simple function $\xi_n^{(l)}$ defined on $[0, \pi]$ for $0 \le n \le N - 1$ by

$$\xi_n^{(l)}(\lambda) = \sqrt{\frac{f(t_1)}{\pi}} \cos(nt_1) \mathbf{1}_{\{0\}}(\lambda) + \sum_{k=0}^{l-1} \sqrt{\frac{f(t_{k+1})}{\pi}} \cos(nt_{k+1}) \mathbf{1}_{(t_k, t_{k+1}]}(\lambda)$$

which is similar to the typical construction of stochastic integral.

Define the sine counterpart as $\theta_n^{(l)}(\lambda)$, and then integrate both $\xi_n^{(l)}(\lambda)$ and $\theta_n^{(l)}(\lambda)$ with respect to $dB_1(\lambda)$ and $dB_2(\lambda)$ on $[0, \pi]$ to approximate X_n. Then, we have

$$\hat{X}_n^{(l)} = \sum_{k=0}^{l-1} \sqrt{\frac{f(t_{k+1})}{l}} \left[\cos(nt_{k+1}) U_k^{(0)} - \sin(nt_{k+1}) U_k^{(1)} \right]$$

where $U_k^{(\cdot)}$ are i.i.d. standard normal random variables. $U_k^{(0)}$ and $U_k^{(1)}$ are independent, as they are resulted from integration from the two aforementioned independent Brownian motions.

Similar to the Fourier transform approach, the fGns can be recovered by applying the FFT to the sequence of $\hat{X}_n^{(l)}$ efficiently to the following coefficient:

$$(4.13) \qquad a_k = \begin{cases} 0 & k = 1 \\ \frac{1}{2}\left(U_{k-1}^{(0)} + iU_{k-1}^{(1)}\right) & k = i, \cdots, l - 1 \\ U_{k-1}^{(0)} \sqrt{\frac{f(t_k)}{l}} & k = l \\ \frac{1}{2}\left(U_{2l-k-1}^{(0)} + iU_{2l-k-1}^{(1)}\right) & k = l + 1, \cdots, 2l - 1 \end{cases}$$

It is easy to check that the covariance structure of fGns can be recovered, with the help of product-to-sum trigonometric identity, as

$$\begin{aligned} Cov\left(\hat{X}_m^{(l)}, \hat{X}_n^{(l)}\right) &= \sum_{k=0}^{l-1} \frac{f(t_{k+1})}{l} \cos((m-n) t_{k+1}) \\ &\cong 2 \int_0^\pi \frac{f(\lambda)}{2\pi} \cos(n\lambda) \, d\lambda \\ &= \frac{1}{2\pi} \int_{-\pi}^\pi f(\lambda) \exp(-in\lambda) \, d\lambda = \gamma(n) \end{aligned}$$

Paxson [33] has also proposed another method for simulating fGns, where in [23] it was proven to be related to (4.13) with the case $l = N/2$:

$$b_k = \begin{cases} 0 & k = 0 \\ \sqrt{\frac{R_k f(t_k)}{N}} \exp(i\Phi_k) & k = 1, \cdots, N/2 - 1 \\ \sqrt{\frac{f(t_{N/2})}{2N}} U_{N/2}^{(0)} & k = N/2 \\ b_k^* & k = N/2 + 1, \cdots, N - 1 \end{cases}$$

Here, R_k is a vector of exponentially distributed random variables with mean 1, and Φ_k are uniformly distributed random variables on $[0, 2\pi]$ independent of R_k. This method is of order $N \log(N)$, and only one FFT is required instead of 2 times compared to the Davis-Harte FFT method. Hence, it is about 4 times faster.

Remark: The Paxson algorithm in (4.13) is improved by Dieker [23] to retain the normality of the sequence and its relationship with the original spectral representation.

5. Numerical Example: fBM Volatility Model

Finally, this section provides a numerical example of Monte Carlo simulation of fBm volatility model. In Section 2.3, we have briefly mentioned the truncated fractional Brownian motion. This section outlines the stochastic volatility model explored by Comte and Renault [14]. We follow the example given in [14], with the following setup to simulate the volatility process:

$$(5.1) \qquad \begin{cases} \sigma(t) = \sigma_0 e^{x(t)} \\ dx(t) = -kx(t)\,dt + v d\widehat{B^H}(t) \end{cases}$$

where v is the volatility factor of the log-volatility process $x(t)$. The volatility process is the exponential of an OU-process driven by the truncated fBm. Also, we assume that $x(0) = 0$, $k > 0$, $1/2 < H < 1$.

Solving the OU-process with integrating factor, we have

$$(5.2) \qquad x(t) = \int_0^t v e^{-k(t-s)} d\widehat{B^H}(s)$$

By applying the fractional calculus or using the formulas provided in [13], we can formulate $x(t)$ in another way as

$$(5.3) \qquad x(t) = \int_0^t a(t-s)\,dB(s)$$

where $B(t)$ is an ordinary standard Brownian motion and

$$(5.4) \qquad \begin{aligned} a(\theta) &= \frac{v}{\Gamma(H+1/2)} \frac{d}{dx} \int_0^\theta e^{-ku}(\theta - u)^{H-1/2}\,du \\ &= \frac{v}{\Gamma(H+1/2)} \left(\theta^\alpha - ke^{-\theta} \int_0^\theta e^{ku} u^\alpha\,du\right) \end{aligned}$$

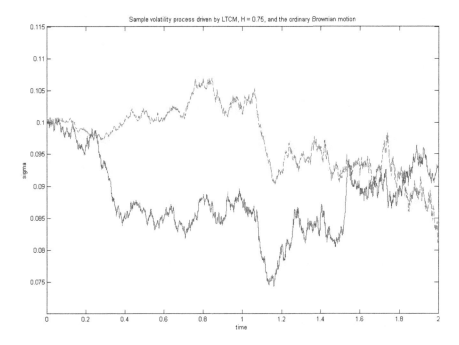

Figure 2. shows a sample path of $\sigma(t) = exp(\tilde{x}(t))$ for $k = 1; \sigma_0 = 0 : 1; v = 0 : 3; H = 0 : 75; T = 2$ (bottom line) is shown here. For the purpose of comparison, a sample path of the volatility process driven by an ordinary OU-process (the line above) is shown alongside, which has the same parameters except the Hurst index equals to 0.5.

By applying the ordinary discretization scheme to (5.3), we have

$$(5.5) \qquad \tilde{x}(t) = \sum_{j=1}^{t_N=t} a\left(t_N - \frac{t_j - 1}{n}\right)\left(B\left(t_j\right) - B\left(t_{j-1}\right)\right)$$

Here, the coefficient $a(\cdot)$ can be calculated by symbolic packages such as matlab and mathematics. In our case of OU-process, it is a summation of constant with incomplete gamma function and gamma function.

The sample path of the fractional-OU driven volatility process has shown more of a persistent trend, i.e. more prominent trend (more smooth and less reversal) compared to the ordinary-OU driven volatility process, which is what to be expected according [13]. Though this approach generates readily available sample path robustly, this is, in a stricter sense of word, not a real fractional Brownian motion, but a truncated process that imitate it closely and asymptotically. For more discussion of its statistical property and justification of its stability as compared to the original stationary version, we direct the reader to [13] and [14]. We provided

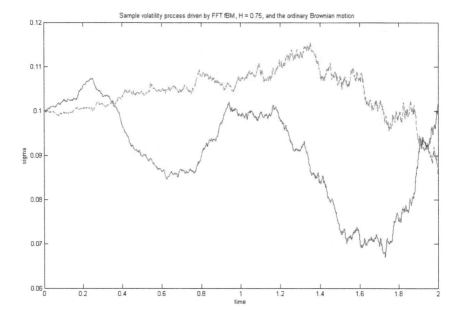

Figure 3. FFT-generated fBM sample path (bottom line) with Hurst index $H = 0.75$, and its ordinary Brownian motion counterpart (top line).

this alternative approach for readers who are more concerned with the practical aspect of such process.

We have also included a 'real' fractional Brownian motion simulated by the circulant-embedding method, with the same parameters as above in Figure 3. Figure 3 shows a sample path of $(t) = \exp(x(t))$, where x(t) is the fBM generated by the circulant-embedding FFT with the same parameters as Figure 2: $k = 1, \sigma_0 = 0.1, v = 0.3, H = 0.75, T = 2$ .In these two examples, the fractional Brownian motions are scaled, so that the variance of $B^H(T)$ equals to the ordinary Brownian motion $B^{1/2}(T)$.

6. Full simulation Scheme

Now we have the robust fractional Brownian motion simulation, the goal is to have a fully functional pricing function with Monte-Carlo simulation. The result will serve as an reference for the following sections.

First we look at the basic setup this:

$$(6.1) \quad \begin{aligned} d \ln S_t &= (r - \tfrac{1}{2}\sigma_t^2)dt + \sigma_t d\tilde{W}_t \\ dX_t &= \kappa_t(\theta - X_t)dt + v_t^H dW_t^H \\ \sigma_t &= \exp(X_t) \\ d < W, \tilde{W} >_t &= \rho_t dt \end{aligned}$$

This setup is the basic setup similar to Comte and Renault's, where the volatility is an exponential of an OU-process. This setup is to capture the mean-reverting nature of the volatility process. Note here we did not restrict the volatility Brownian motion to be a fractional Brownain motion or not (i.e. we did not specify the hurst index besides that $\frac{1}{2} \leq H < 1$). The reason will become apparent later. The W_t is called the "pre-transformed fBM" for the volatility process. The idea behind this process is imple, we want to impose correlation between asset's processs and volatility's process, naturally we want impose it between the asset brownian motion and the driving force behind the stochastic volatility. The problem here is that, the covariation of brownian motions with different hurst indices converge to zero. So it is naturally to seek another way, one of the most intuitive way is to impose correlation with the brownian motion that is used as seed in the simulation of the brownian motion. Looking back at the section that involves the the FFT approach of the fractional Brownian motion transform, we see that it involves a seed that is the standard normal distribution, it will be quite simple to impose a linear correlation of the ordinary brownian motion generated by this standard normal distribution and the asset process's ordinary brownian motion. And that's indeed what we will do here.

Another way is of course generate another path by FFT with $H = \frac{1}{2}$, and then impose correlation with methods such as Cholesky theorem. We will outline the reason why this is implemented later.

6.1 Conditional Distribution

Since we impose the exponential-OU process structure, it is easy to see that the volatility has a log-normal distribution, from the probability distribution function where the long-term mean is

$$e^{\theta+0.5Var(X_T)}$$

In order to arrive at a particular long-term mean, it is necessary to calculate this value explicitly, it is quite involved for a fractional brownian motion drive process X_T, we outline the calculation put forth by Pipiras and Taqqu (2001), in the following sections it is not necessary to do so since the parameteres, thus the long-term mean is calibrated against the market data instead of having a hard-set target. We included the methodolgy here for completion sake.

6.1.1 Fractional Riemann-Liouville Integral

In order to understand the theoractical calculation involves the fractional brownian motion, it is necessary to introduce the notion of the fractional integral, following the work of Pipiras and Taqqu, we choose the Riemann-Liouville integral, which is defined in the compact interval $[0, T]$, for some $T > 0$, a finite time horizon and in order to be consistent with the paper we will be citing, we have the following Hurst index notation: $\kappa = H - \frac{1}{2}$, and we are only interested in the case $\kappa > 0$, displaying long-range dependence. For some $f \in L^2(\mathbb{R})$, we have the Riemann-Liouville integral

Definition 6.1.

$$(I^\kappa_{T-}f)(s) = \frac{1}{\Gamma(\kappa)} \int_s^T f(r)(r-s)^{\kappa-1}dr, \qquad 0 \le s \le T$$

This integral that always exists $f \in L^2(\mathbb{R})$ this always exists.

And it has the fractional derivative which is the counterpart of the fractional integral

$$(D^\kappa_{T-}g)(u) = -\frac{1}{\Gamma(1-\kappa)}\left(\frac{g(u)}{(T-u)^\kappa} + \kappa \int_u^T \frac{g(u)-g(s)}{(s-u)^{\kappa+1}}ds\right), \qquad 0 < u < T$$

Similar to ordinary calculus, we have the following equality: $I^{-\kappa}_{T-} = D^\kappa_{T-}$.

For $\kappa = 0$, $I^\kappa_{T-} = D^\kappa_{T-} = id$.

Theorem 6.1. (fractional Brownian motion integrand space) *The following space of possible integrand has been introduced for $\kappa \in (0, \frac{1}{2})$:*

$$\Lambda^\kappa_T := \left\{ f : [0, T] \to \mathbb{R} \,\Big|\, \int_0^T \left[s^{-\kappa}I^\kappa_{T-}\left((\cdot)^\kappa f(\cdot)\right)(s)\right]^2 ds < \infty \right\}$$

for $f, g \in \Lambda^\kappa_T$, define the scalar product

$$\langle f, g \rangle_{\kappa,T} := \frac{\pi\kappa(2\kappa+1)}{\Gamma(1-2\kappa)\sin(\pi\kappa)} \int_0^T s^{-2\kappa}\left[I^\kappa_{T-}\left((\cdot)^\kappa f(\cdot)\right)(s)\right]\left[I^\kappa_{T-}\left((\cdot)^\kappa g(\cdot)\right)(s)\right]$$

for $\kappa = 0, \langle f, g \rangle_{\kappa,T} = \langle f, g \rangle_{L^2}$, and $\langle \cdot, \cdot \rangle_{\kappa,T} = \|\cdot\|_{\kappa,T}$

With the fractional stochastic integral, we have the following isometry:

$$\left\|\int_0^T c(s)dB^\kappa(s)\right\|_2^2 = \|c(\cdot)\|_{\kappa,T}$$

We have the following lemma for conditional expectation:

Lemma 6.1.

$$E\left[B^\kappa(t)|B^\kappa(s), v \in [0, s]\right] = B^\kappa(s) + \int_0^s \Psi^\kappa(s, t, v)dB^\kappa(v)$$

Where for $v \in (0, t)$,

$$\Psi(s, t, v) = \frac{\sin(\pi\kappa)}{\pi}v^{-\kappa}(s-v)^{-\kappa}\int_s^t \Psi^\kappa(s, t, v)dB^\kappa(v)$$

Then for $c \in \Lambda^\kappa_T$,

$$E\left[\int_0^t c(v)dB^\kappa(v)\Big| B^\kappa(v), v \in [0, s]\right] = \int_0^s c(v)dB^\kappa(v) + \int_0^s \Psi^\kappa_c(s, t, v)dB^\kappa(v)$$

For such an fractionally stochastic integral, we have the following characteristic function:

Lemma 6.2.

$$E\left[e^{iu\int_0^t c(v)dB(v)}\Big|\mathcal{F}_s\right] = \exp\left\{iu\left[\int_0^s c(v)dB^\kappa(v) + \int_0^s \Psi_c^\kappa(s,t,v)dB^\kappa(v)\right]\right\}$$

$$\times \exp\left\{-\frac{u^2}{2}\left\|c(\cdot)\mathbf{1}_{[s,t]}(\cdot)\right\|_{\kappa,T}^2 - \left\|\Psi_c^\kappa(s,t,\cdot)\mathbf{1}_{[0,s]}(\cdot)\right\|_{\kappa,T}^2\right\}$$

So the fractionally stochastic integral $\int_0^t c(u)dB^\kappa(u)|\mathcal{F}_s$ is normally distributed with the following moments:

$$E\left[\int_0^t c(v)dB^\kappa(v)|\mathcal{F}_s\right] = \int_0^s c(v)dB^\kappa(v) + \int_0^s \Psi_c^\kappa(s,t,v)dB^\kappa(v)$$

$$Var\left[\int_0^t c(v)dB^\kappa(v)|\mathcal{F}_s\right] = \left\|c(\cdot)\mathbf{1}_{[s,t]}(\cdot)\right\|_{\kappa,T}^2 - \left\|\Psi_c^\kappa(s,t,\cdot)\mathbf{1}_{[0,s]}(\cdot)\right\|_{\kappa,T}^2$$

With the OU-process driven by fBM, Fink, Kluppelberg, Zahle (2010), has proved the following:

Lemma 6.3.

$$dX(t) = (k(t) - a(t)X(t))\,dt + \sigma(t)dB^\kappa(t), \quad X(0) \in \mathbb{R}, \quad t \in [0,T]$$

$$X(t) = X(0)e^{-\int_0^t a(s)ds} + \int_0^t e^{-\int_s^t a(u)du}k(s)ds + \int_0^t e^{-\int_s^t a(u)du}\sigma(s)dB^\kappa(s), \; t \in [0,T]$$

Then similar to the previous characteristic function, we have the following:

Lemma 6.4.

$$E\left[e^{iuX(t)}\Big|\mathcal{F}_s\right] = \exp\left\{iu\left[X(s)e^{-\int_s^t a(v)dv} + \int_s^t e^{-\int_v^t a(w)dw}k(v)dv + \int_0^s \Psi_c^\kappa(s,t,v)dB^\kappa(v)\right]\right\}$$

$$\times \exp\left\{-\frac{u^2}{2}\left\|c(\cdot)\mathbf{1}_{[s,t]}(\cdot)\right\|_{\kappa,T}^2 - \left\|\Psi_c^\kappa(s,t,\cdot)\mathbf{1}_{[0,s]}(\cdot)\right\|_{\kappa,T}^2\right\}$$

with $c(\cdot) = \sigma(\cdot)e^{-\int_\cdot^t a(w)dw}$, $X(t)|\mathcal{F}_s$ is normally distributed with

$$E\left[X(t)|\mathcal{F}_s\right] = X(s)e^{-\int_s^t a(v)dv} + \int_s^t e^{-\int_v^t a(w)dw}k(v)dv + \int_0^s \Psi_c^\kappa(s,t,v)dB^\kappa(v)$$

$$Var\left[X(t)|\mathcal{F}_s\right] = \left\|c(\cdot)\mathbf{1}_{[s,t]}(\cdot)\right\|_{\kappa,T}^2 - \left\|\Psi_c^\kappa(s,t,\cdot)\mathbf{1}_{[0,s]}(\cdot)\right\|_{\kappa,T}^2$$

The difficulty of calculating this quantity lies within calculating the norm of the fractional integral, which as we have seen, has a singularity at $t = \dot{0}$. Fink, Kluppelberg, Zahle (2010) has outlined a discretization scheme for $\kappa \in (0, \frac{1}{2})$, which I will included here:

Theorem 6.2.

$$\left\| c(\cdot) \mathbf{1}_{[0,t]}(\cdot) \right\|_{\kappa,T}^2 = \frac{\pi\kappa(2\kappa+1)}{\Gamma(1-2\kappa)\sin(\pi\kappa)(\Gamma(\kappa))^2} \int_0^T s^{-2\kappa} \left(\int_s^T \frac{r^\kappa c(r) \mathbf{1}_{[0,t]}(r)}{(r-s)^{1-\kappa}} dr \right)^2 ds$$

The next step is to discretize the integral, decomposing the outer integral, $n \in \mathbb{N}$, and $0 = s_0 \le s_1 \le \cdots \le s_n = T$.

$$\int_0^T s^{-2\kappa} \left(\int_s^T \frac{r^\kappa c(r) \mathbf{1}_{[0,t]}(r)}{(r-s)^{1-\kappa}} dr \right)^2 ds = \sum_{i=0}^{n-1} \int_{s_i}^{s_{i+1}} s^{-2\kappa} \left(\int_s^T \frac{r^\kappa c(r) \mathbf{1}_{[0,t]}(r)}{(r-s)^{1-\kappa}} dr \right)^2 ds$$

Within the $s \in [s_i, s_{i+1}]$, we can approximate by extending the lower limit:

$$\int_s^T \frac{r^\kappa c(r) \mathbf{1}_{[0,t]}(r)}{(r-s)^{1-\kappa}} dr \approx \int_{s_i}^T \frac{r^\kappa c(r) \mathbf{1}_{[0,t]}(r)}{(r-s_i)^{1-\kappa}} dr$$

For $i = 0, \cdots, n-1$, and partition $[s_i, s_{i+1}]$ into $s_i = u_0^i \le u_1^i \le \cdots \le u_{m_i}^i = s_{i+1}$, for some $m_i \in \mathbb{N}$.

$$\int_{s_i}^T \frac{r^\kappa c(r) \mathbf{1}_{[0,t]}(r)}{(r-s_i)^{1-\kappa}} dr = \sum_{j=0}^{m_i-1} \int_{u_j^i}^{u_{j+1}^i} \frac{r^\kappa c(r) \mathbf{1}_{[0,t]}(r)}{(r-s_i)^{1-\kappa}} dr$$

$$\approx \frac{1}{\kappa} \sum_{j=0}^{m_i-1} \left[\left(u_{j+1}^i - s_i \right)^\kappa - \left(u_j^i - s_i \right)^\kappa \right] \frac{\left(u_j^i \right)^\kappa c(u_j^i) + \left(u_{j+1}^i \right)^\kappa c\left(u_{j+1}^i \right)}{2}$$

Since we have $\Gamma(\kappa) \cdot \kappa = \Gamma(\kappa+1)$,

$$\left\| c(\cdot) \mathbf{1}_{[0,t]}(\cdot) \right\|_{\kappa,T}^2 = \frac{\pi\kappa(2\kappa+1)}{\Gamma(2-2\kappa)\sin(\pi\kappa)(2\Gamma(\kappa+1))^2} \sum_{i=0}^{n-1} \left[s_{i+1}^{1-2\kappa} - s_i^{1-2\kappa} \right]$$

$$\times \left[\sum_{j=0}^{m_i-1} \left[\left(u_{j+1}^i - s_i \right)^\kappa - \left(u_j^i - s_i \right)^\kappa \right] \left(u_j^i \right)^\kappa c\left(u_j^i \right) + \left(u_{j+1}^i \right)^\kappa c\left(u_{j+1}^i \right) \right]^2$$

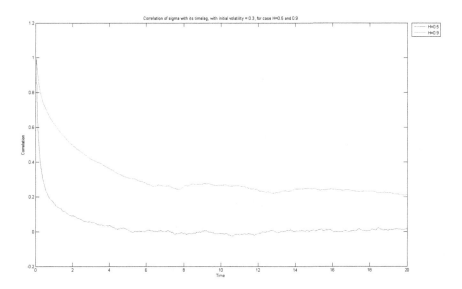

Figure 4. Correlation of σ (Exponential OU Process).

Choose $s_i = 0.01, i = 0, \cdots, 100t$ *and* $u_j^i = 0.01 (i + j), j = 0, \cdots, 100t - i,$
we have

$$\left\| c\left(\cdot\right) \mathbf{1}_{[0,t]}\left(\cdot\right) \right\|_{\kappa,T}^2$$

$$\approx \frac{\pi\kappa\left(2\kappa + 1\right)}{\Gamma\left(2 - 2\kappa\right) \sin\left(\pi\kappa\right)\left(2\Gamma\left(\kappa + 1\right)\right)^2} 0.01^{1+2\kappa} \sum_{i=0}^{100t-1} \left(\left[\left(i + 1\right)^{1-2\kappa} - i^{1-2\kappa}\right]\right.$$

$$\left. \times \sum_{j=0}^{100t-i-1} \left[\left(j + 1\right)^{\kappa} - j^{\kappa}\right]\left[\left(i + j\right)^{\kappa} c\left(0.01\left(i + j\right)\right) + \left(i + j + 1\right)^{\kappa} c\left(0.01\left(i + j + 1\right)\right)\right]\right)$$

For proof and detail we refer reader to the Pipiras and Taqqu paper, and the discretization scheme and the proof to Fink, Kluppelberg, Zahle (2010).

6.2 Full simulation - Stochastic Volatility

6.2.1 Uncorrelated Cases

Here we outline the simulation scheme as previously stated, the stochastic volatility part of the process is implemented by FFT-generated fBM. This is to captured the volatility persistence phenomenon found in long-maturity option markets. The ordinary stochastic volatility (such as Heston model) cannot sufficiently captured this, as the volatility term-structure decays much faster than expected in the ordinary stochastic volatility model, and fractional brownian motion random source provides a easy and direct enrichment of the existing and well-developed framework. One of the most direct way to observe volatility persistence is to ob-

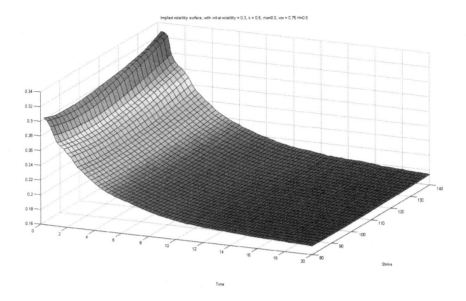

Figure 5. Implied Volatility Surface, $\rho = 0, v = 0.75, H = 0.5$.

serve the mean-volatility and the correlation structure, as we have pointed out earlier, volatility persistence displays the behavior of long-range dependence, which can be observed by looknig at the autocovariance structure. In figure 4 depicts the correlation between the sigma and its time-lagged counterparts (unit is displayed in years), where the initial volatility is 0.3, and only difference of the two plots are the hurst index being 0.5 (ordinary brownian motion) and 0.9 (fractional brownian motion). The Long-range dependence can be seen from the plot, becuase while the ordinary brownian motion case (H=0.5) quickly dwindled close to zero, the fractional brownian motion case (H=0.9) stays signifantly above zero. So obviously, for the first case the quantity $\sum_{n=1}^{\infty} \gamma(n)$ converges to a number while the 2nd quantity it diverges and satisfy the requirement of a long-range dependence.

For the case of full stochastic volatility simulation, the european option (call and put) have been calculated and then the implied volatility is plotted. The setup is as follow (unless otherwise specified): $T = 20, \sigma_0 = 0.3, \kappa_t = 0.5, v_t^H = 0.75$.

The first graph is the implied volatility surface where the stochastic volaitlity is generated by ordinary brownian motion. Second graph is the same setup only that the hurst index = 0.9, the last graph is combined for comparasion.

From the figure 5 to 7: first, the volatility for the case of fractional Brownian motion driven stochastic volatility persists much longer than the ordinary brownian motion driven one. Also on the volatility surface, the out-of-money ends

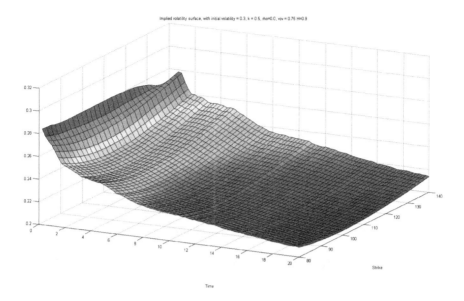

Figure 6. Implied Volatility Surface, $\rho = 0, \nu = 0.75, H = 0.9$.

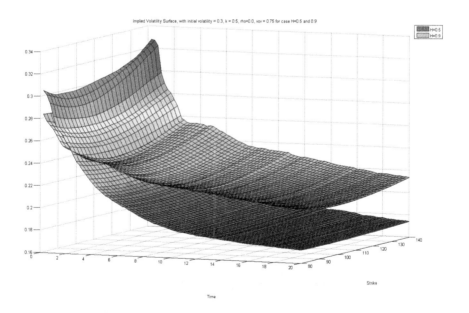

Figure 7. Combined Implied Volatility Surface, $\rho = 0, \nu = 0.75, H = 0.5, 0.9$.

of the close-to-maturity display some kind of numerical instability. This is what we aimed for in the first place, obviously for market data with a high long-term maturity volatility level, the ordinary brownian motion can adapt this by turning up the vol-of-vol, but this is defeating the purpose. Case-in-point, for example, if one wants to model a distribution that displays fat-tailed behavior, one can of course fit a normal-distribution on the data, but it will produce a normal distribution with a high variance that seems like an 'overkill', obviously it is mismodelling. Simialr story can be told for our case, where it will be much easier to adapt to the slow-decaying volatility surface by implementing the fractionally-driven stochastic volaitlity process instead of the ordinary stochastic volatility with a much higher vol-of-vol than necessary.

6.2.2 Correlated Case, skewness

It is quite intuitive to impose correlation between the volatility and the asset processes to attempt to display some kind of volatility skewness. It was first proposed to simulate the log-volatility process with the fractional brownian motion, where the "pre-transformed fBM", i.e. the Brownian motion, generating seed of the stochastic volatility fractional Brownian motion, is to be imposed with a linear correlation with the asset's own Brownian motion, similar to the case of the ordinary stochastic volatility setup. This is similar to the idea of constructing Brownian motion with fractional integration. Or can be written succintly in terms of M-operator in S-transform introduced by Bender (2003) and Oksendal, where we introduce briefly here:

Definition 6.2.

$$\widehat{Mf}(y) = |y|^{1/2-H}\,\widehat{f}(y), \qquad y \in \mathbb{R}$$

where

$$\widehat{g}(y) := \int_{\mathbb{R}} e^{-ixy} g(x)\, dx$$

denotes the Fourier transform.

For $0 < H < 1$

$$Mf(x) = -\frac{d}{dx}\frac{C_H}{(H-1/2)} \int_{\mathbb{R}} (t-x)|t-x|^{H-3/2} f(t)\, dt$$

Where

$$C_H = \left\{ 2\Gamma\left(H-\frac{1}{2}\right)\cos\left[\frac{\pi}{2}\left(H-\frac{1}{2}\right)\right] \right\}^{-1} [\Gamma(2H+1)\sin(\pi H)]^{\frac{1}{2}}$$

Here $\Gamma(\cdot)$ denotes the classical gamma function.

For the case we are interested in:

Definition 6.3. For $\frac{1}{2} < H < 1$

$$Mf(x) = C_H \int_{\mathbb{R}} \frac{f(t)}{|t - x|^{3/2-H}} dt$$

And if $H = \frac{1}{2}$, then

$$Mf(x) = f(x)$$

Also

$$\|f\|_H := \|Mf\|_{L^2(\mathbb{R})}$$

And the inner product on this space is:

$$\langle f, g \rangle_H = \langle Mf, Mg \rangle_{L^2(\mathbb{R})}$$

Define

$$M\mathbf{1}_{[0,t]}(x) := M[0,t] x$$

For the indicator function $\mathbf{1}_{[a,b]}(x)$, for $a < b$

$$M[a,b](x) = \frac{[\Gamma(2H+1)\sin(\pi H)]^{1/2}}{2\Gamma(H+1/2)\cos[\pi/2(H+1/2)]} \left[\frac{b-x}{|b-x|^{3/2-H}} - \frac{a-x}{|a-x|^{3/2-H}} \right]$$

This was calculated by applying the direct definition of the M-operator on the indicator function.

Theorem 6.3.
   With Parseval's Theorem, we know

$$\int_{\mathbb{R}} [M[a,b](x)]^2 = \frac{1}{2\pi} \int_{\mathbb{R}} \left[\widehat{M[a,b]}(\xi) \right]^2 d\xi$$

$$= \frac{1}{2\pi} \int_{\mathbb{R}} |\xi|^{1-2H} \frac{|e^{-ib\xi} - e^{-ia\xi}|^2}{|\xi|^2} d\xi$$

$$= (b-a)^{2H}$$

Here we have applied the fact:

$$\widehat{I[a,b]}(\xi) = \frac{\left[e^{-ib\xi} - e^{-ia\xi} \right]}{-i\xi}$$

Since $M[s,t] = M[0,t] - M[0,s]$ for $s < t$, by polar-identity

$$\int_{\mathbb{R}} M[0,t](x) M[0,s](x) dx = \frac{1}{2} \left(|t|^{2H} + |s|^{2H} - |t-s|^{2H} \right)$$

And this coincides with the autocovariance structure from the beginning of the paper.

Theorem 6.4. *Furthermore, from the paper it is shown that:*

$$E\left[B^H(s)B^H(t)\right] = \int_{\mathbb{R}} M[0,s](x)M[0,t](x)\,dx$$

$$= \frac{1}{2}\left(|t|^{2H} + |s|^{2H} - |s-t|^{2H}\right)$$

Finally we have the definition:

Definition 6.4.

$$\int_{\mathbb{R}} f(t)\,dB^H(t) = \int_{\mathbb{R}} Mf(t)\,dB(t), \qquad f \in L_H^2(\mathbb{R})$$

It's easy to see:

$$B^H(t) = \int_{\mathbb{R}} M_H[0,t](u)\,dB(u)$$

Here $M_H[\cdot,\cdot]$ is the M-operator with Hurst-index H,

Conversely, we have:

Lemma 6.5.

$$B(t) = \int_{\mathbb{R}} M_{1-H}[0,t](u)\,dB^H(u)$$

Remark 6.1. Actually, the M-operator can be succintly written in terms of fractional integration:

$$M_H f = K_H I_-^{H-1/2} f, \qquad \text{for } \frac{1}{2} < H < 1$$

Where $I_-^{H-1/2}$ is a fractional integration to the order of $H - 1/2$. This definition coincide with Comte, Renault's if the process is defined only on $[0,t]$.

With Definition 6.4, Alos (2007) has proved that for the case of $H > \frac{1}{2}$ with Malliavin Calculus that, the skew-slope of the implied volatility around the At-The-Money strike close to maturity $\tau \to 0$ does not depend on the correlation imposed between the asset's Brownian motion and stochastic volatility's "pre-transformed fBM". This is a rather surprisingly finding, which we will investigate with our numerical simulation.

Similar to previous sub-section, we have the similar setup, and just to check for validity of the claim of Alos, we impose the correlation in the 'pre-transformed fBM' for the case where the correlation is non-zero.

One can see by comparing the figure 8 and 9 that, the skewness is very pronouced in the ordinary brownian motion, but in the case of the fractional brownian motion, the correlation does not have any effect on the skewness of the implied volatility, this is indeed a very interesting finding, which agree with the

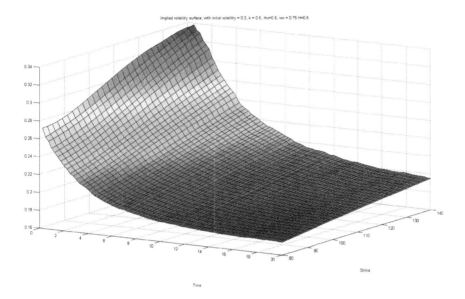

Figure 8. Implied Volatility Surface, $\rho = 0.5, \nu = 0.75, H = 0.5$.

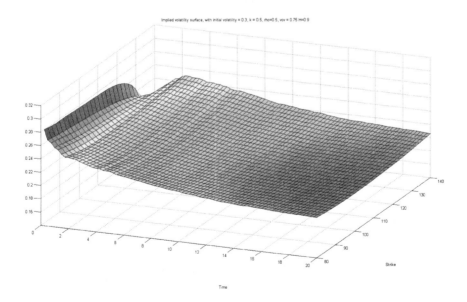

Figure 9. Implied Volatility Surface, $\rho = 0.5, \nu = 0.75, H = 0.9$.

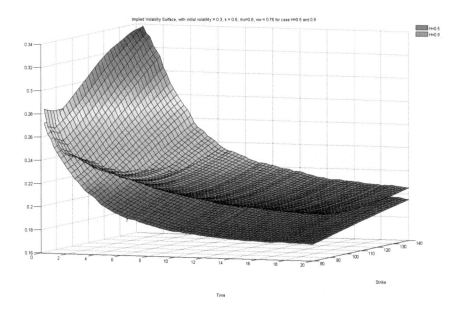

Figure 10. Combined Implied Volatility Surface, $\rho = 0.5, \nu = 0.75, H = 0.5, 0.9$.

theorem by Alos, and one of the main result of the paper. Also the fractional Brownian motion driven stochastic volatility also display a slower decay of the volatilty surface compared to the ordinary Brownian motion version.

6.3 Mixture fBM exponential volatility

Instead of choosing between the ordinary Brownian motion and fractional Brownian motion, in order to capture both the skewness and volatility persistence, both Brownian motion and fractional Brownian motion are utilized in order. The formal setup of the model is outlined as following:

(6.2)
$$d \ln S_t = \left(r_t - \frac{1}{2}\sigma_t^2 \right) dt + \sigma_t d\widetilde{W}_t$$
$$dX_t = \kappa_t \left(\theta - X_t \right) dt + v_t dW_t + v_t^H dW_t^H$$
$$\sigma_t = \exp(X_t)$$
$$d \left\langle W, \widetilde{W} \right\rangle_t = \rho_t d_t$$

The implied volatility surface of this mixture fBM exponential volatility setup is shown in figure (11), with parameter $\rho = 0.5, v_t = 0.5, v_t^H = 0.4, H = 0.9$.

There are some drawbacks of this formulation: Large vol-of-vol might lead to big jump in volatility, and causes numerical instability, this is due to the exponential form of the volatility that magnifies the large deviation from the fractional Brownian motion and results in very slow convergence. There might be special

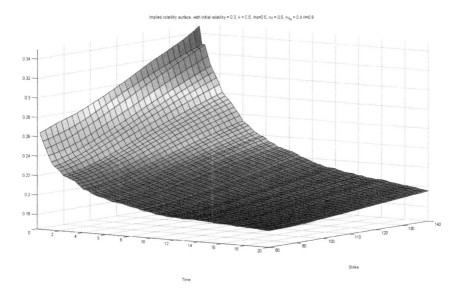

Figure 11. Mixed fBM exponential volatility Implied Volatility Surface, $\rho = 0.5$, $\nu_t = 0.5$, $v_t^H = 0.4$, $H = 0.9$.

difficulty to calibrate due to high similarity of fractional Brownian Motion and ordinary Brownian motion.

7. Conclusion

Motivated by the inadequacy in capturing long-dependence feature in volatility process of the traditional stochastic volatility framework, we explore the possibility of fractional Brownian motion (fBm) in financial modeling and various schemes of the Monte Carlo simulation. Starting from the general definition, fBm can be considered as an extension of the ordinary Brownian motion with an autocovariance function that depends on both time indices instead of just the minimum between the two. With different values of Hurst index, we can distinguish fractional Brownian motion into three different cases: $H < 1/2, H = 1/2$ and $H > 1/2$. Since only the case of $H > 1/2$ displays a long-dependence behavior, that is the case we are interested in. Several prominent examples of fBm in financial modeling are given. Simulation schemes are divided into the exact schemes and approximate schemes. While the former will capture the complete structure for the whole length of sample size, the latter either approximates the value of the real realization or truncates the covariance structure for robustness. We start with the exact scheme of Hosking algorithm that utilizes the multivariate Gaussian distribution of fractional Gaussian noises and simulates the sample points conditioned on the previous samples. Alternatively, instead of simulating each conditioned on the past sample points, we can first construct the covariance matrix of the size of the sample we want, and proceed to find the 'square root' of the covariance matrix and multiply with a standard normal variable vector, for which the product vector will be the fractional Gaussian noise (fGn) with exact covariance structure as the covariance matrix. To find the 'square root', we first investigate the Cholesky decomposition, but the computational and memory expense is too large to be feasible in practice. In contrast, fast Fourier transform (FFT) simulation embeds the original covariance matrix in a larger circulant matrix and simulates by diagonalizing the circulant matrix into the product of eigenvalue matrix and unitary matrix. The FFT method is significantly more robust than the previous schemes. We then look into the approximate schemes; namely the construction of fBm with correlated random walks, which can be viewed as an extension of construction of Brownian motion with ordinary random walk. This method gives us interesting insight into the true working of fBm, especially the idea of long-range dependence. This approach is not only interesting and easy to implement, but also the error can be calculated explicitly. The drawback of this approach is that the speed slows down significantly with large sample points, and the tradeoff is made based on the error function. The last simulation approach we look at is the conditional random midpoint displacement (RMD) scheme, which is mathematically similar to the Hosking scheme, but with fixed number of past sample points it conditions on. The on-the-fly version of RMD scheme can indefinitely generate sample points with given resolution. Finally, we include the spectral method for approximating fBm. Comparing all the schemes and also referring the studies done in [32], we conclude that if the time-horizon is known beforehand, the FFT/spectral schemes would be the best scheme due to the high speed and accuracy. Alternately, if samples should be generated indefinitely, the on-the-fly conditioned RMD scheme seems to offer similar level of accuracy and speed as the FFT scheme. At last, instead of the usual definition of fBm as in [2], we also look at the truncated version of fBm proposed in [13], [14], which is generally not equivalent to the fBm though it retains some important features such as self-similarity and long-range dependence for $H > 1/2$. The volatility process that is driven by a fractional-OU process is included as an example. First we made a discovery through full simulation of the fBM driven Stochastic volatility process and found out that, even if we impose correlation between the asset and volatility process, it would not affect the skewness of the implied volatility surface, this is further support by E.Alos's work by Malliavin Calculus in the paper [34]. So it motivates us to come up with the mixture-fBM model, which has both the ordinary Brownian motion and the fractional Brownian motion in the stochastic volatility process, and is able to capture volatility persistence, volatility smile and skewness.

References

1. A. Kolmogorov, "Wienersche spiralen und einige andere interessante kurvenim hilbertschen raum," *C.R. (Doklady) Acad. URSS (N.S)*, vol. 26, pp. 115–118, 1940.
2. B. Mandelbrot and J. Van Ness, "Fractional brownian motions, fractional noises and applications," *SIAM Review*, vol. 10, pp. 422–437, 1968.
3. J. Hull and A. White, "The pricing of options on assets with stochastic volatilities," *Rev. Finan. Studies*, vol. 3, pp. 281–300, 1987.
4. L. Scott, "Option pricing when the variance changes randomly: Estimation and an application," *J. Financial and Quant. Anal.*, vol. 22, pp. 419–438, 1987.
5. Z. Ding, C. Granger, and R. Engle, "A long memory property of stock market returns and a new model," *J. Empirical Finance*, vol. 1, pp. 83–106, 1993.
6. N. Crato and P. de Lima, "Long-range dependence in the conditional variance of stock returns," *Economics Letters*, vol. 45, pp. 281–285, 1994.
7. R. Baillie, T. Bollerslev, and H. Mikkelsen, "Fractionally integrated generalized autoregressive conditional heteroskedasticity," *J. Econometrics*, vol. 74, pp. 3–30, 1996.
8. T. Bollerslev and H. Mikkelsen, "Modeling and pricing long memory in stock market volatility," *J. Econometrics*, vol. 73, pp. 151–184, 1996.
9. E. Jacquier, N. Polson, and P. Rossi, "Bayesian analysis of stochastic volatility models," *J. Bus. and Econ. Statistics*, vol. 12, pp. 371–417, 1994.
10. F. Breidt, N. Crato, and P. de Lima, "The detection and estimation of long memory in stochastic volatility," *J. Econometrics*, vol. 83, pp. 325–348, 1998.
11. F. Biagini, Y. Hu, B. Oksendal, and T. Zhang, *Stochastic Calculus for Fractional Brownian Motion and Applications*. Springer, 2008.
12. R. Cont, "Volatility clustering in financial markets: Empirical facts and agent-based models," in *Long Memory in Economics*, pp. 289–310, Springer, 2007.
13. F. Comte and E. Renault, "Long memory continuous time models," *J. Econometrics*, vol. 73, pp. 101–149, 1996.
14. F. Comte and E. Renault, "Long memory in continuous-time stochastic volatility models," *Mathematical Finance*, vol. 8, pp. 291–323, 1998.
15. D. Marinucci and P. Robinson, "Alternative forms of fractional brownian motion," *J. Statistical Planning and Inference*, vol. 80, pp. 111–122, 1999.
16. F. Delbaen and W. Schachermayer, "A general version of the fundamental theorem of asset pricing," *Math. Ann.*, vol. 300, pp. 463–520, 1994.
17. L. Rogers, "Arbitrage with fractional brownian motion," *Mathematical Finance,*, vol. 7, pp. 95–105, 1997.
18. P. Cheridito, "Arbitrage in fractional brownian motion models," *Finance and Stochastics*, vol. 7, pp. 533–553, 2003.
19. F. Comte, L. Coutin, and E. Renault, "Affine fractional stochastic volatility models," *Annals of Finance*, vol. 8, pp. 337–378, 2012.
20. D. Duffie, R. Pan, and K. Singleton, "Transform analysis and asset pricing for affine jump-diffusion," *Econometrica*, vol. 68, pp. 1343–1376, 2000.
21. M. Fukasawa, "Asymptotic analysis for stochastic volatility: Martingale expansion," *Finance and Stochastics*, vol. 15, pp. 635–654, 2011.
22. N. Yoshida, "Malliavin calculus and martingale expansion," *Bull. Sci. Math.*, vol. 125, pp. 431–456, 2001.
23. T. Dieker, *Simulation of Fractional Brownian Motion*. Master thesis, Vrije Univer-

siteit, Amsterdam, 2002.

24. J. Hosking, "Modeling persistence in hydrological time series using fractional differencing," *Water Resources Research*, vol. 20, pp. 1898–1908, 1984.

25. P. Brockwell and R. Davis, *Time Series: Theory and Methods*. New York: Springer, 1987.

26. R. Davis and D. Harte, "Tests for hurst effect," *Biometrika,*, vol. 74, pp. 95–102, 1987.

27. C. Dietrich and G. Newsam, "Fast and exact simulation of stationary gaussian processes through circulant embedding of the covariance matrix," *SIAM J. Sci. Computing*, vol. 18, pp. 1088–1107, 1997.

28. E. Perrin, R. Harba, R. Jennane, and I. Iribarren, "Fast and exact synthesis for 1-d fractional brownian motion and fractional gaussian noises," *IEEE Signal Processing Letters*, vol. 9, pp. 382–384, 2002.

29. A. Wood and G. Chan, "Simulation of stationary gaussian processes in [0,1]^d," *J. Computational and Graphical Statistics*, vol. 3, pp. 409–432, 1994.

30. N. Enriquez, "A simple construction of the fractional brownian motion," *Stochastic Processes and Their Applications*, vol. 109, pp. 203–223, 2004.

31. M. Taqqu, "Weak convergence to fractional brownian motion and to the rosenblatt process," *Z. Wahr. Verw. Gebiete*, vol. 31, pp. 287–302, 1975.

32. I. Norros, P. Mannersalo, and J. Wang, "Simulation of fractional brownian motion with conditionalized random midpoint displacement," *Advances in Performance Analysis*, vol. 2, pp. 77–101, 1999.

33. V. Paxson, "Fast, approximate synthesis of fractional gaussian noise for generating self-similar network traffic," *Computer Communication Review*, vol. 27, pp. 5–18, 1997.

34. E. Alos, J. Leon, and J. Vives, "On the short-time behavior of the implied volatility for jump-diffusion models with stochastic volatility," *Finance and Stochastics*, vol. 11, no. 4, pp. 571–589, 2007.

Mean-Variance Pre-Commitment Policies Revisited Via a Mean-Field Technique

[1]A. Bensoussan,[*] [2]K. C. Wong,[†] [3]S. C. P. Yam[‡]

[1]International Center for Decision and Risk Analysis, School of Management, The University of Texas at Dallas
Graduate Department of Financial Engineering, Ajou University
Department of Systems Engineering and Engineering Management, City University of Hong Kong
[2]Department of Mathematics, The University of Hong Kong
[3]Department of Statistics, The Chinese University of Hong Kong

In this article, we provide an alternative approach, based on mean-field theory, on establishing the pre-commitment optimal solution to the continuous time mean-variance portfolio selection problem. By considering the probability distribution of the underlying wealth process and introducing a suitable adjoint function, we show that the optimal control (pre-commitment solution) satisfies a HJB equation that also involves the probability distribution of the corresponding optimal wealth process, which naturally satisfies Kolmogorov's Forward Equation. For if the probability density function exists, it will also satisfy a Fokker-Planck (Kolmogorov's Backward Equation), and hence together with the mentioned HJB, the whole problem becomes a special problem in the theory of mean-field games. Solving by *Ansatz*, we further obtain the explicit optimal asset allocation, which coincides with that first obtained in the work of Li and Zhou (2000) [18].

1. Introduction

Since the introduction by Markowitz (1952) [20], the mean-variance portfolio selection problem has become one of key research topics in finance. The investor aims to determine the optimal portfolio, which minimizes the risk, measured by

---

[*]E-mail: axb046100@utdallas.edu
[†]E-mail: wongkc89@hku.hk
[‡]Email: scpyam@sta.cuhk.edu.hk

the variance of the terminal wealth, subject to a predetermined budget constraint and at an arbitrary level of expected terminal return. Later on, Merton (1971) [21] extended the Markowitz's model to the continuous time setting, he investigated both the optimal consumption and portfolio selection problems by formulating them as a stochastic control problem.

Many researchers use utility function to assess the performance of portfolio, instead of using the expected return and variance of wealth directly, because dynamic programming can be applied with the convenience of Tower Property, for instance, see Merton (1971) [21], Pliska (1986) [24]. On the other hand, the nonlinear nature of the variance makes Tower Property fail to hold, and so results that neither dynamic programming method nor Bellman optimality principle can be adopted. In other words, an optimal control that optimizes the mean-variance utility at time zero needs not to remain to be optimal for the mean-variance utility at any latter time. In literature, this phenomenon is commonly referred as *time inconsistency*. One direction of resolution on these time inconsistent problem is to pre-commit at an initial time, namely: by finding the optimal control for objective function at the initial time; whether or not this optimal control would serve as optimal for the same objective function at a latter time is not considered. For examples, in Li and Zhou (2000) [18] and Chen et al. (2008) [11], their obtained optimal controls depend on both the initial and current states. Due to the non-optimal nature, in the "future" of the planning horizon, of pre-commitment approach, some researchers propose an alternative time consistent approach originated from the notion of "subgame perfect Nash equilibria", see Peleg and Yaari (1973) [25], Björk and Murgoci [6], Bensoussan et al. [5]; and also see Bensoussan et al. [4] in the mean-field game setting. Note that this time consistent strategy admits Bellman optimality principle, and hence it could be sought out by tackling a system of (extended) HJB equations.

Solving for the precommitment solution to the mean-variance problem is usually difficult because of the failure of Bellman optimality principle. By applying embedding technique of solving for the portfolio selection problem under mean-variance criterion directly, Li and Zhou (2000) [18] and Li and Ng [17] provide ground-breaking works on tackling the problem in continuous time and multi-period settings respectively, in which they converted the original mean-variance problem into an auxiliary stochastic linear-quadratic problem. The same embedding technique has been widely used for tackling general mean-variance optimization problems, to name a few: (1) Lim and Zhou (2002) [19] considered the mean-variance problem with random parameter; (2) Yin and Zhou (2003) [29] considered the problem with regime switching modulated parameters; (3) Chiu and Li (2006) [12] considered mean-variance asset-liability management problems; (4) Chen et al. (2008) [11] included both regime switching parameters and asset-liability features into the mean-variance problem. However, the optimal control under pre-commitment is time inconsistent and hence is difficult to be sought out; yet it is still practically meaningful, as the pre-commitment control always outperforms the time-consistent counterpart, e.g. see Wei et al. (2013) [27].

In this article, we utilize the mean-field technique to solve for the pre-commitment solution to the mean-variance portfolio selection problem, it is obviously different from the approach in Li and Zhou (2000) [18], in which they adopted the mentioned embedding technique. Mean-field game and mean-field type control theory has recently become a popular mathematical research topic in engineering, economic and financial, and probabilistic and statistical context with a wide range of applications; see Bensoussan et al. (2013) [2] and references therein. To motivate the notion of mean-field games naturally arisen in economic and financial problems, due to the dramatic population growth and rapid urbanization, urgent needs of in-depth understanding of collective strategic interactive behaviors of a huge group of investors is crucial to maintaining sustainable economic growth. To model their interactions without much computational cost, all interactions can be comprehended by a single mean-field term, which originates from statistical physics. And mathematically, this mean-field term is essentially a functional of the equilibrium population density. For some recent results in the theory of mean-field game problems, one can consult the works, such as Huang et al. (2007) [9], Lasry and Lions (2007) [16], Bensoussan et al. (2011) [3], and Bensoussan et al. (2012) [4]. In general, to solve for a mean-field game problem, we first seek for the optimal control by using the maximum principle with respect to an arbitrary (but fixed) dynamics z in the place of the mean-field term. We next consider the optimal dynamics y with respect to the previously obtained optimal control and z, and then look for a fixed point of the map $z \mapsto \mathbb{E}(y)$, or some other similar expected functionals of y. This general methodology can be found in Bensoussan et al. (2011) [3] and Bensoussan et al. (2013) [2]. On the other hand, for optimal control problems of mean-field type, they can be regarded as those classical control problems so that both the system dynamics and the corresponding objective functionals may involve certain "mathematical expectations" (mean-field terms) of the present state of the system dynamics. For some recent results in the mean-field type control theory, one may consult Andersson and Djehiche (2011) [1], Buckdahn et al. (2011) [8], Meyer-Brandis et al. (2011) [22], Yong (2011) [28], and Nourian (2012) [10].

For our present mean-variance portfolio selection problem, we transform it into a mean-field problem as follows. In principle, we first set the probability distribution of wealth process, with respect to an arbitrary control, to satisfy the Kolmogorov's Forward Equation (Fokker-Planck equation if the density exists); we next derive the optimality condition (HJB equation) based on maximum principle so that the obtained HJB equation involves the mean-field term in terms of the probability distribution of the corresponding optimal wealth, with respect to the optimal control that could be established as the fixed point by solving this HJB equation together with the Kolmogorov's Forward Equation (resp. Fokker-Planck equation).

In Section 2, we introduce our problem setting. In Section 3, we show that the optimal control satisfies a HJB equation that involves a mean-field term; in the subsection, we further discuss about the sufficient condition leading to the existence and smoothness of the transition density of optimal wealth process. In Section 4, we derive the optimal asset allocation by considering a suitable form of the solution to HJB equation as obtained in Section 3. Finally, in Section 5, we shall show that our optimal control is the same as that in Li and Zhou (2000) [18].

2. Problem Setting

Let $(\Omega, \mathcal{F}, \mathbb{P})$ be a fixed complete probability space, over which $W(t) = (W_1(t), \ldots, W_d(t))'$ denotes a d-dimensional standard Brownian motion, M' denote the transpose of a matrix M.

Suppose that the market has $m+1$ assets with m risky assets with price process $S(t) = (S_1(t), \ldots, S_m(t))'$ and one riskless money account with price process $B(t)$ such that the pair $(B(t), S(t))$ satisfies the following equations respectively

$$dB(t) = r(t)B(t)dt,$$

$$dS_k(t) = \mu_k(t)S_k(t)\,dt + S_k(t)\sum_{j=1}^{d} \sigma_{kj}(t)dW_j(t),$$

where $r(t)$ is the riskless interest-rate, $\mu_k(t)$ and $\sigma_k(t) \triangleq (\sigma_{k1}(t), \ldots, \sigma_{kd}(t))$ are respectively the appreciation rate and volatility processes of the k-th risky asset, and these are all assumed to be bounded. Also define a volatility matrix of assets as $\sigma(t) = \left(\sigma_{kj}(t)\right)_{m \times d}$, we assume its non-degeneracy so that $\sigma(t)\sigma(t)' \geq \delta I$ for some $\delta > 0$, and therefore $(\sigma(t)\sigma(t)')^{-1}$ exists.

Let $u_k(t)$ be the money amount invested in the k-th risky asset of the portfolio at time t. We consider the class of admissible controls to be the family of all $\mathcal{L}^2$-integrable feedback control $u(t, x) \triangleq (u_1(t, x), \ldots, u_m(t, x))'$ (which may also depend on x_0), that is

$$\|u\|^2 := \mathbb{E}\left[\int_0^T u(t, X^u(t))' u(t, X^u(t))dt\right] < \infty,$$

such that $u(x, t)$ also satisfies (i) the following Lipschitz continuity:

(1) $\qquad |u(t, x) - u(t, y)| \leq K^u |x - y|, \quad$ for all $(t, x, y) \in [0, T] \times \mathbb{R}^2,$

and (ii) the linear growth condition:

(2) $$\sup_{(t,x)} \frac{|u(t, x)|}{1 + |x|} < K^u,$$

for some constant $K^u < \infty$. Here, $|x| := \sqrt{x_1^2 + \cdots + x_n^2}$ for any $x = (x_1, \ldots, x_n)' \in \mathbb{R}^n$.

The dynamics of a controlled wealth process is:

$$dX^u(t) = r(t)X^u(t)dt + u(t, X^u(t))'(\alpha(t)dt + \sigma(t)dW(t))$$

(3)
$$= (r(t)X^u(t) + u(t, X^u(t))'\alpha(t))dt + u(t, X^u(t))'\sigma(t)dW(t)$$

where $\alpha(t) \triangleq (\alpha_1(t), \ldots, \alpha_m(t))'$ and $\alpha_k(t) \triangleq \mu_k(t) - r(t)$ for $k \in \{1, \ldots, m\}$. We also let the initial wealth to be $X_0 = x_0$. Condition (1) guarantees the existence of solution of the SDE (3). Define an objective functional:

(4)
$$J(u) := \mathbb{E}[X^u(T)] - \gamma(x_0)Var[X^u(T)],$$

where $T < \infty$ and risk aversion coeffiecient $\gamma(x)$ is assumed to be positive for all $x \in \mathbb{R}$. The present article aims to establish an admissible feedback control u so that it maximizes $J(u)$.

3. Euler Optimality Conditions

Under our consideration here, each control, given the initial state, could be regarded as "Markovian", and hence so are the corresponding controlled wealth processes. Denote the transition probability measure, with respect to a feasible controul u, by $m^u(\cdot, t)$ (i.e. $\mathbb{P}(X^u(t) < y) = \int_{-\infty}^{y} m^u(dx, t)$). In accordance with Conditions (1) and (2), by applying Theorem 1.2.1 in Stroock (2008) [26], we have $m^u(\cdot, t)$ continuous (with respect to weak* topology) in t and satisfying the following Kolmogorov Forward equation:

For any $\Phi \in C^2(\mathbb{R})$,

$$\int_{-\infty}^{\infty} \Phi(x)m^u(dx, t) - \int_{-\infty}^{\infty} \Phi(x)m^u(dx, 0)$$
$$= \int_{0}^{t} \int_{-\infty}^{\infty} \left[[r(s)x + u(s, x)'\alpha(s)]\frac{d\Phi}{dx}(x) + \frac{1}{2}u(s, x)'\sigma(s)\sigma(s)'u(s, x)\frac{d^2\Phi}{dx^2}(x) \right] m^u(dx, s)ds.$$

(5)

On the other hand, since $Var(X(t)) = \mathbb{E}[X(t)^2] - (\mathbb{E}[X(t)])^2$, so

$$J(u) = \int_{-\infty}^{\infty} (x - \gamma(x_0)x^2)m^u(dx, T) + \gamma(x_0)\left(\int_{-\infty}^{\infty} xm^u(dx, T) \right)^2.$$

Given a feasible control $\hat{u}$ and an arbitrary admissible control u and $\theta > 0$, denote $\hat{m} = m^{\hat{u}}$ and $\Delta^\theta m = m^{\hat{u}+\theta u} - m^{\hat{u}}$, we then have

$$J(\hat{u} + \theta u) - J(\hat{u}) = \int_{-\infty}^{\infty} (x - \gamma(x_0)x^2)\Delta^\theta m(dx, T)$$
$$+2\gamma(x_0)\int_{-\infty}^{\infty} x\Delta^\theta m(dx, T)\int_{-\infty}^{\infty} x\hat{m}(dx, T) + \gamma(x_0)\left(\int_{-\infty}^{\infty} x\Delta^\theta m(dx, T) \right)^2.$$

Before we proceed on , we first give an estimate result that X^u is $\mathcal{L}^2$-integrable if u is also $\mathcal{L}^2$-integrable. Define $r := \max_{t \in [0,T]} r(t)$, $\alpha := \max_{t \in [0,T]} |\alpha(t)|$ and $\sigma := \max_{t \in [0,T]} \sum_{i,j} \sigma_{ij}(t)^2$.

Lemma 3.1. *Given that u is $\mathcal{L}^2$-integrable, then*

$$\mathbb{E}\left[\int_0^T (X^u(t))^2 \, dt \right] < \infty.$$

Proof. By Itô's formula, we have

$$
\begin{aligned}
&d\,(X^u(t))^2 \\
&= \left\{ 2r(t)\,(X^u(t))^2 + 2X^u(t)\alpha(t)'u(t, X^u(t)) + u(t, X^u(t))'\sigma(t)\sigma(t)'u(t, X^u(t)) \right\} dt \\
&\quad + 2X^u(t)u(t, X^u(t))'\sigma(t)dW(t) \\
&\leq \left\{ (2r(t) + 1)\,(X^u(t))^2 + u(t, X^u(t))' \left[\sigma(t)\sigma(t)' + \alpha(t)\alpha(t)' \right] u(t, X^u(t)) \right\} dt \\
&\quad + 2X^u(t)u(t, X^u(t))'\sigma(t)dW(t) \\
&\leq \left\{ (2r(t) + 1)\,(X^u(t))^2 + (\sigma^2 + \alpha^2)u(t, X^u(t))'u(t, X^u(t)) \right\} dt + 2X^u(t)u(t, X^u(t))'\sigma(t)dW(t),
\end{aligned}
$$

The first inequality follows from the fact that $2ab \leq a^2 + b^2$ for any $a, b \in \mathbb{R}$. Taking expectation on both sides, we have

$$d\mathbb{E}\left[(X^u(t))^2 \right] \leq \left\{ (2r(t) + 1)\mathbb{E}\left[(X^u(t))^2 \right] + (\sigma^2 + \alpha^2)\mathbb{E}\left[u(t, X^u(t))'u(t, X^u(t)) \right] \right\} dt.$$

By using Gronwall's inequality, we have

$$\mathbb{E}\left[(X^u(t))^2 \right] \leq (\sigma^2 + \alpha^2)e^{\int_0^t (2r(s)+1)ds} \int_0^t \mathbb{E}\left[u(s, X^u(s))'u(s, X^u(s)) \right] ds$$

$$\leq (\sigma^2 + \alpha^2)e^{\int_0^t (2r(s)+1)ds} \|u\|^2 .$$

In particular,

$$\mathbb{E}\left[\int_0^T (X^u(t))^2 \, dt \right] \leq \left(\int_0^T (\sigma^2 + \alpha^2)e^{\int_0^t (2r(s)+1)ds} dt \right) \|u\|^2 .$$

Our claim follows since $r(t)$ is bounded uniformly. $\square$

Next, we also have the following preliminary results for certain moment estimations:

Lemma 3.2. *For any $0 \leq t \leq T$, we have*

1.

$$\mathbb{E}\left[|X^{\hat{u}+\theta u}(t) - X^{\hat{u}}(t)| \right] \leq K\theta,$$

2.

$$\mathbb{E}\left[|X^{\hat{u}+\theta u}(t) - X^{\hat{u}}(t)|^2\right] \le K\theta^2,$$

where K is independent of t.

Remark 3.1. The first assertion in Lemma 3.2 implies that $\left|\mathbb{E}[f(X^{\hat{u}+\theta u}(t))] - \mathbb{E}[f(X^{\hat{u}}(t))]\right| \le K'\theta$ for any Lipschitz continuous function f and K' is another constant depending on f. Hence, $m^{\hat{u}+\theta u}(\cdot, t)$ converges weakly to $m^{\hat{u}}(\cdot, t)$ as $\theta \to 0$.

Proof. [Proof of Lemma 3.2] The first assertion follows from the second assertion by a simple application of Jensen's inequality. We now establish the second assertion.

By applying Itô's formula, Cauchy-Schwarz inequality, linearly growth condition on u and Lipschitz continuity of $\hat{u}$, we have:

$$
\begin{aligned}
&d\left(X^{\hat{u}+\theta u}(t) - X^{\hat{u}}(t)\right)^2 \\
&= \Big\{ 2r(t)\left(X^{\hat{u}+\theta u}(t) - X^{\hat{u}}(t)\right)^2 \\
&\quad + 2(1+\theta)\left(X^{\hat{u}+\theta u}(t) - X^{\hat{u}}(t)\right)\alpha(t)'\left(\hat{u}(t, X^{\hat{u}+\theta u}(t)) - \hat{u}(t, X^{\hat{u}}(t))\right) \\
&\quad + 2\theta\left(X^{\hat{u}+\theta u}(t) - X^{\hat{u}}(t)\right)\alpha(t)'u(t, X^{\hat{u}}(t)) \\
&\quad + \Big[(1+\theta)\left(\hat{u}(t, X^{\hat{u}+\theta u}(t))' - \hat{u}(t, X^{\hat{u}}(t))'\right) + \theta u(t, X^{\hat{u}}(t))'\Big]\sigma(t)\sigma(t)' \\
&\quad \cdot \Big[(1+\theta)\left(\hat{u}(t, X^{\hat{u}+\theta u}(t)) - \hat{u}(t, X^{\hat{u}}(t))\right) + \theta u(t, X^{\hat{u}}(t))\Big]\Big\} dt \\
&\quad + 2\left(X^{\hat{u}+\theta u}(t) - X^{\hat{u}}(t)\right)\left((\hat{u}+\theta u)(t, X^{\hat{u}+\theta u}(t)) - \hat{u}(t, X^{\hat{u}}(t))\right)'\sigma(t)\,dW(t) \\
&\le \Big\{\Big[2r + 2\alpha K^{\hat{u}}(1+\theta) + 1\Big]\left(X^{\hat{u}+\theta u}(t) - X^{\hat{u}}(t)\right)^2 + \theta^2 u(t, X^{\hat{u}}(t))'\alpha(t)\alpha(t)'u(t, X^{\hat{u}}(t)) \\
&\quad + 2\sigma^2\Big[(1+\theta)^2\left(\hat{u}(t, X^{\hat{u}+\theta u}(t))' - \hat{u}(t, X^{\hat{u}}(t))'\right)\left(\hat{u}(t, X^{\hat{u}+\theta u}(t)) - \hat{u}(t, X^{\hat{u}}(t))\right) \\
&\quad + \theta^2 u(t, X^{\hat{u}}(t))'u(t, X^{\hat{u}}(t))\Big]\Big\} dt + \{\cdots\}\,dW(t) \\
&\le \Big\{\Big[2r + 2\alpha K^{\hat{u}}(1+\theta) + 1 + 2(1+\theta)^2\sigma^2(K^{\hat{u}})^2\Big]\left(X^{\hat{u}+\theta u}(t) - X^{\hat{u}}(t)\right)^2 \\
&\quad + 2\theta^2(\sigma^2 + \alpha^2)u(t, X^{\hat{u}}(t))'u(t, X^{\hat{u}}(t))\Big\} dt + \{\cdots\}\,dW(t) \\
&\le \Big\{\Big[2r + 2\alpha K^{\hat{u}}(1+\theta) + 1 + 2(1+\theta)^2\sigma^2(K^{\hat{u}})^2\Big]\left(X^{\hat{u}+\theta u}(t) - X^{\hat{u}}(t)\right)^2 \\
&\quad + 2\theta^2(\sigma^2 + \alpha^2)(K^u)^2(1 + |X^{\hat{u}}(t)|)^2\Big\} dt + \{\cdots\}\,dW(t).
\end{aligned}
$$

Applying expectation and then integration from 0 to t on both sides of the last inequality, we deduce that

$$\mathbb{E}\left[\left(X^{\hat{u}+\theta u}(t) - X^{\hat{u}}(t)\right)^2\right]$$

$$\leq \int_0^t \left[2r + 2\alpha K^{\hat{u}}(1+\theta) + 1 + 2(1+\theta)^2\sigma^2(K^{\hat{u}})^2\right]\mathbb{E}\left[\left(X^{\hat{u}+\theta u}(s) - X^{\hat{u}}(s)\right)^2\right]dt$$

$$+ \int_0^t 4\theta^2(\sigma^2 + \alpha^2)(K^u)^2\mathbb{E}\left[1 + |X^{\hat{u}}(s)|^2\right]ds.$$

By using Gronwall's inequality, we can see that the second assertion follows:

$$\mathbb{E}\left[(X^{\hat{u}+\theta u}(t) - X^{\hat{u}}(t))^2\right]$$

$$\leq 4\theta^2(\sigma^2 + \alpha^2)(K^u)^2 e^{[2r+2\alpha K^{\hat{u}}(1+\theta)+1+2(1+\theta)^2\sigma^2(K^{\hat{u}})^2]t}\left(t + \int_0^t \mathbb{E}|X^{\hat{u}}(s)|^2 ds\right)$$

$$\leq 4\theta^2(\sigma^2 + \alpha^2)(K^u)^2 e^{[2r+2\alpha K^{\hat{u}}(1+\theta)+1+2(1+\theta)^2\sigma^2(K^{\hat{u}})^2]T}\left(T + \int_0^T \mathbb{E}|X^{\hat{u}}(s)|^2 ds\right). \quad \Box$$

Define an adjoint function $\phi \in C^{2,1}$ which satisfies the following duality equation:

$$(6) \quad -\frac{\partial\phi(x,t)}{\partial t} - (r(t)x + \hat{u}(t,x)'\alpha(t))\frac{\partial\phi(x,t)}{\partial x} - \frac{1}{2}\hat{u}(t,x)'\sigma(t)\sigma(t)'\hat{u}(t,x)\frac{\partial^2\phi(x,t)}{\partial x^2} = 0$$

where $\phi(x,T) = x - \gamma(x_0)x^2 + 2\gamma(x_0)x \int_{-\infty}^{\infty} \xi\hat{m}(d\xi,T).$

Assumption 3.1. *For technical reasons, we assume that the following two conditions could be satisfied by the adjoint function ϕ:*

1. $\frac{\partial^2\phi}{\partial x^2} < 0$

2. *There exists a function $g(x)$ so that*

$$\left|\frac{\partial\phi}{\partial t}(x,t)\right| \leq g(x), \forall x \quad and \quad \mathbb{E}\left[\int_0^T g(X_t^u)\right] < \infty \quad for\ every\ admissible\ control\ u.$$

(7)

By Lemma 3.2 (1), $\int_{-\infty}^{\infty} x\Delta^\theta m(dx,T) = O(\theta)$, thus we have

$$J(\hat{u} + \theta u) - J(\hat{u}) = \int_{-\infty}^{\infty}\left(x - \gamma(x_0)x^2 + 2\gamma(x_0)x \int_{-\infty}^{\infty} \xi\hat{m}(d\xi,T)\right)\Delta^\theta m(dx,T) + O(\theta^2)$$

$$= \int_{-\infty}^{\infty} \phi(x,T)\Delta^\theta m(dx,T) + O(\theta^2)$$

For any large enough $N \in \mathbb{N}$, we discretize the time horizon into $[0, T]$ into N subintervals $[t_k, t_{k+1}]$ of equal length, where $t_k := \frac{kT}{N}$ for $k = 0, 1, \ldots, N$. Since $\Delta^\theta m(dx, 0) = 0$, we then have

$$J(\hat{u} + \theta u) - J(\hat{u}) = \sum_{k=1}^{N} \int_{-\infty}^{\infty} \left\{ \phi(x, t_k) \Delta^\theta m(dx, t_k) - \phi(x, t_{k-1}) \Delta^\theta m(dx, t_{k-1}) \right\} + O(\theta^2)$$

$$= \sum_{k=1}^{N} \int_{-\infty}^{\infty} \left\{ \phi(x, t_k) [\Delta^\theta m(dx, t_k) - \Delta^\theta m(dx, t_{k-1})] \right\}$$

$$(8) \qquad + \sum_{k=1}^{N} \int_{-\infty}^{\infty} \left\{ [\phi(x, t_k) - \phi(x, t_{k-1})] \Delta^\theta m(dx, t_{k-1}) \right\} + O(\theta^2).$$

Since m is continuous in t, a simple application of Fubini's theorem together with Assumption 3.1 (2) on ϕ implies that

$$(9)$$

$$\lim_{N \to \infty} \sum_{k=1}^{N} \int_{-\infty}^{\infty} [\phi(x, t_k) - \phi(x, t_{k-1})] \Delta^\theta m(dx, t_{k-1}) = \int_0^T \int_{-\infty}^{\infty} \frac{\partial \phi}{\partial t}(x, t) \Delta^\theta m(dx, t) dt.$$

On the other hand, apply the forward equation (5) to each $\phi(\cdot, t_k)$, we have

$$\sum_{k=1}^{N} \int_{-\infty}^{\infty} \left\{ \phi(x, t_k) [\Delta^\theta m(dx, t_k) - \Delta^\theta m(dx, t_{k-1})] \right\}$$

$$= \sum_{k=1}^{N} \int_{t_{k-1}}^{t_k} \left\{ \int_{-\infty}^{\infty} \left[[r(s)x + (\hat{u}(s, x) + \theta u(s, x))'\alpha(s)] \frac{\partial \phi}{\partial x}(x, t_k) \right. \right.$$

$$+ \frac{1}{2}(\hat{u}(s, x) + \theta u(s, x))'\sigma(s)\sigma(s)'(\hat{u}(s, x) + \theta u(s, x)) \frac{\partial^2 \phi}{\partial x^2}(x, t_k) \right] m^{\hat{u}+\theta u}(dx, s)$$

$$- \int_{-\infty}^{\infty} \left[[r(s)x + \hat{u}(s, x)'\alpha(s)] \frac{\partial \phi}{\partial x}(x, t_k) \right.$$

$$\left. \left. + \frac{1}{2}\hat{u}(s, x)'\sigma(s)\sigma(s)'\hat{u}(s, x) \frac{\partial^2 \phi}{\partial x^2}(x, t_k) \right] \hat{m}(dx, s) \right\} ds$$

$$= \sum_{k=1}^{N} \int_{t_{k-1}}^{t_k} \int_{-\infty}^{\infty} \left[[r(s)x + \hat{u}(s, x)'\alpha(s)] \frac{\partial \phi}{\partial x}(x, t_k) \right.$$

$$\left. + \frac{1}{2}\hat{u}(s, x)'\sigma(s)\sigma(s)'\hat{u}(s, x) \frac{\partial^2 \phi}{\partial x^2}(x, t_k) \right] \Delta^\theta m(dx, s) ds$$

$$+ \sum_{k=1}^{N} \int_{t_{k-1}}^{t_k} \int_{-\infty}^{\infty} u(s, x)' \left(\alpha(s) \frac{\partial \phi}{\partial x}(x, t_k) + \sigma(s)\sigma(s)'\hat{u}(s, x) \frac{\partial^2 \phi}{\partial x^2}(x, t_k) \right) m^{\hat{u}+\theta u}(dx, s) ds$$

$$+ O(\theta^2).$$

By the smoothness of ϕ and taking limit as N goes to infinity on both sides, we obtain

$$\lim_{N\to\infty} \sum_{k=1}^{N} \int_{-\infty}^{\infty} \left\{ \phi(x, t_k)[\Delta^\theta m(dx, t_k) - \Delta^\theta m(dx, t_{k-1})] \right\}$$

$$= \int_0^T \int_{-\infty}^{\infty} \left[[r(s)x + \hat{u}(s, x)'\alpha(s)]\frac{\partial\phi}{\partial x}(x, s) + \frac{1}{2}\hat{u}(s, x)'\sigma(s)\sigma(s)'\hat{u}(s, x)\frac{\partial^2\phi}{\partial x^2}(x, s) \right] \Delta^\theta m(dx, s)ds$$

$$+ \theta \int_0^T \int_{-\infty}^{\infty} u(s, x)' \left(\alpha(s)\frac{\partial\phi}{\partial x}(x, s) + \sigma(s)\sigma(s)'\hat{u}(s, x)\frac{\partial^2\phi}{\partial x^2}(x, s) \right) m^{\hat{u}+\theta u}(dx, s)ds + O(\theta^2).$$

(10)

Substitute (9) and (10) into (8) and apply the definition of adjunct function ϕ, we further obtain:

$$J(\hat{u} + \theta u) - J(\hat{u})$$

$$= \int_0^T \int_{-\infty}^{\infty} \frac{\partial\phi}{\partial t}(x, t)\Delta^\theta m(dx, s)ds$$

$$+ \int_0^T \int_{-\infty}^{\infty} \left[[r(s)x + \hat{u}(s, x)'\alpha(s)]\frac{\partial\phi}{\partial x}(x, s) + \frac{1}{2}\hat{u}(s, x)'\sigma(s)\sigma(s)'\hat{u}(s, x)\frac{\partial^2\phi}{\partial x^2}(x, s) \right] \Delta^\theta m(dx, s)ds$$

$$+ \theta \int_0^T \int_{-\infty}^{\infty} u(s, x)' \left(\alpha(s)\frac{\partial\phi}{\partial x}(x, s) + \sigma(s)\sigma(s)'\hat{u}(s, x)\frac{\partial^2\phi}{\partial x^2}(x, s) \right) m^{\hat{u}+\theta u}(dx, s)ds + O(\theta^2)$$

$$= \theta \int_0^T \int_{-\infty}^{\infty} u(s, x)' \left(\alpha(s)\frac{\partial\phi}{\partial x}(x, s) + \sigma(s)\sigma(s)'\hat{u}(s, x)\frac{\partial^2\phi}{\partial x^2}(x, s) \right) m^{\hat{u}+\theta u}(dx, s)ds + O(\theta^2).$$

(11)

To derive the necessary result in relation to a HJB equation, we first need the following lemma:

Lemma 3.3. *Under Assumption 3.1 (1), i.e. $\frac{\partial^2\phi}{\partial x^2} \leq 0$, for every $(t, x) \in [0, T] \times \mathbb{R}$, the following two statements are equivalent:*

1. *$\hat{u}(t, x)$ maximize*

$$\frac{\partial\phi(x, t)}{\partial x}\alpha(t)'u + \frac{1}{2}\frac{\partial^2\phi(x, t)}{\partial x^2}u'\sigma(t)\sigma(t)'u \quad among\ all\ u \in \mathbb{R}^m.$$

2. *$\hat{u}(t, x)$ satisfies*

$$(u(t, x)' - \hat{u}(t, x)')\left(\alpha(t)\frac{\partial\phi(x, t)}{\partial x} + \sigma(t)\sigma(t)'\hat{u}(t, x)\frac{\partial^2\phi(x, t)}{\partial x^2} \right) \leq 0, \ for\ all\ u(t, x) \in \mathbb{R}^m.$$

Proof. Our claim follows as an immediate consequence of the Kuhn-Tucker's condition. $\square$

Note that our estimate (11) holds for any admissible controls u and $\hat{u}$, we have the following necessary condition for optimality.

Theorem 3.1. *Given an admissible control $\hat{u}$, and ϕ is defined by (6) with $\hat{m} = m^{\hat{u}}$. If $\hat{u}$ is the optimal control and ϕ satisfies Assumption 3.1, the following HJB equation holds on the support of $\hat{m}(\cdot, t)$:*

$$-\frac{\partial\phi(x,t)}{\partial t} - r(t)x\frac{\partial\phi(x,t)}{\partial x} = \max_{u(t,x)}\left[\frac{\partial\phi(x,t)}{\partial x}\alpha(t)'u(t,x) + \frac{1}{2}\frac{\partial^2\phi(x,t)}{\partial x^2}u(t,x)'\sigma(t)\sigma(t)'u(t,x)\right],$$

(12)

where $\qquad \phi(x,T) = x - \gamma(x_0)x^2 + 2\gamma(x_0)x\int_{-\infty}^{\infty}\xi\hat{m}(d\xi, T),$

for every $(t,x) \in [0,T] \times \mathbb{R}$ such that $\hat{m}(dx,t) > 0$.

Remark 3.2. Recall that the measure $\hat{m} := m^{\hat{u}}$ satisfies the Kolmogorov's Forward Equation: for any $\Phi \in C^2(\mathbb{R})$,

$$\int_{-\infty}^{\infty}\Phi(x)\hat{m}(dx,t) - \int_{-\infty}^{\infty}\Phi(x)\hat{m}(dx,0)$$
$$= \int_0^t\int_{-\infty}^{\infty}\left[[r(s)x + \hat{u}(s,x)'\alpha(s)]\frac{d\Phi}{dx}(x) + \frac{1}{2}u(s,x)'\sigma(s)\sigma(s)'\hat{u}(s,x)\frac{d^2\Phi}{dx^2}(x)\right]\hat{m}(dx,s)ds.$$

(13)

Proof. [Proof of Theorem 3.1] By using the estimate (11), for any admissible controls u and $\hat{u}$:

(14)
$$\frac{J(\hat{u} + \theta(u - \hat{u})) - J(\hat{u})}{\theta}$$
$$= \int_0^T\int_{-\infty}^{\infty}(u(s,x)' - \hat{u}(s,x)')\left(\alpha(s)\frac{\partial\phi}{\partial x}(x,s) + \sigma(s)\sigma(s)'\hat{u}(s,x)\frac{\partial^2\phi}{\partial x^2}(x,s)\right)m^{\hat{u}+\theta(u-\hat{u})}(dx,s)ds$$
$$+ O(\theta),$$

By Remark 3.1, $m^{\hat{u}+\theta(u-\hat{u})}(\cdot, s)$ converges weakly to $m^{\hat{u}}(\cdot, s)$, if we take limit on both sides of (14), we have

$$\lim_{\theta\to 0}\frac{J(\hat{u} + \theta(u - \hat{u})) - J(\hat{u})}{\theta}$$
$$= \int_0^T\int_{-\infty}^{\infty}(u(s,x)' - \hat{u}(s,x)')\left(\alpha(s)\frac{\partial\phi}{\partial x}(x,s) + \sigma(s)\sigma(s)'\hat{u}(s,x)\frac{\partial^2\phi}{\partial x^2}(x,s)\right)\hat{m}(dx,s)ds$$

Using the optimality of $\hat{u}$, we clearly have that

$$\lim_{\theta \to 0} \frac{J(\hat{u} + \theta(u - \hat{u})) - J(\hat{u})}{\theta} \leq 0.$$

Since u is arbitrary, we must have

$$(15) \qquad (u(t,x)' - \hat{u}(t,x)') \left(\alpha(t) \frac{\partial \phi(x,t)}{\partial x} + \sigma(t)\sigma(t)' \hat{u}(t,x) \frac{\partial^2 \phi(x,t)}{\partial x^2} \right) \leq 0,$$

for each $(t,x) \in [0,T] \times \mathbb{R}$ such that $\hat{m}(dx,t) > 0$. In accordance with Lemma 3.3, (15) is equivalent to the statement that $\hat{u}$ maximizes the RHS of (12). Since ϕ satisfies (6), (12) follows whenever $\hat{m}(dx,t) > 0$. □

Note that, since the optimal control $\hat{u}$ is the maximizer of the RHS of (12), thus

$$(16) \qquad \hat{u}(x,t) = -(\sigma(t)\sigma(t)')^{-1} \frac{\frac{\partial \phi(x,t)}{\partial x}}{\frac{\partial^2 \phi(x,t)}{\partial x^2}} \alpha(t).$$

Therefore, equation (12) becomes:

$$(17) \qquad -\frac{\partial \phi(x,t)}{\partial t} - r(t)x \frac{\partial \phi(x,t)}{\partial x} = -\frac{1}{2}\alpha(t)'(\sigma(t)\sigma(t)')^{-1}\alpha(t) \frac{\left(\frac{\partial \phi(x,t)}{\partial x} \right)^2}{\frac{\partial^2 \phi(x,t)}{\partial x^2}}$$

$$\phi(x,T) = x - \gamma(x_0)x^2 + 2\gamma(x_0)x \int_{-\infty}^{\infty} \hat{m}(\xi,T)\xi d\xi,$$

with $\hat{m}(dx,t) > 0$. In the rest of this section, we first discuss the issue if the density of $\hat{m}$ exists. In the next section, we aim to establish ϕ satisfying (17) for all (x,t) in $[0,T] \times \mathbb{R}$ with $\hat{m} = m^{\hat{u}}$ and $\hat{u}$ in the form of (16).

3.1 Optimality Condition under the Existence of a Smooth Density of $\hat{m}$

In this subsection, we aim to show that, if the transition density of $\hat{m}$ exists, the solution of our original optimization problem can be obtained by solving for the solution of a couple of HJB and Fokker-Plank equations (and hence in the sense as a mean-field problem). We first provide the revelant results and conditions on which the existence and regularity of the density could be ensured.

For any admissible control u, the correspondng wealth process $X^u(t)$ is a continuous Markov process. We next introduce a common useful condition on u, under which the smooth density of $X^u(t)$ could be guaranteed, namely the Hörmander condition. In particular, Hörmander's theorem (see e.g. [7], [14], [13], [23] and [15]) states that if X_t satisfies Hörmander's condition, X_t should have a smooth transition density. For the ease of the development of our present work, the special case of Hörmander's condition under the one-dimensional setting is given as follows.

Condition 3.1. *[Special Hörmander Condition]* Given a SDE:

$$(18) \qquad dX_t = V_0(t, X_t)dt + \sum_{i=1}^{d} V_i(t, X_t) \circ dW_i,$$

which is in Stratonovich form. For real-valued smooth functions $U(t, x)$ and $V(t, x)$, define the Lie bracket $[U, V](t, x)$ as another smooth function:

$$[U, V](t, x) := \frac{\partial U}{\partial x}(t, x)V(t, x) - \frac{\partial V}{\partial x}(t, x)U(t, x).$$

Define the collection of real numbers $\mathcal{V}_k$ for each $k \in \{0, 1, 2, \ldots\}$ by:

$$\mathcal{V}_0 := \{V_i : i > 0\}, \qquad \mathcal{V}_{k+1} := \left\{[U, V_j] : U \in \mathcal{V}_k, j \geq 0\right\}.$$

We say that the above SDE (18) satisfies the (one-dimensional) Hörmander's condition if for every $(t, x) \in [0, T] \times \mathbb{R}$, there exists some finite $k \geq 0$ such that $\mathcal{V}_k$ contains a non-zero real number.

The following lemma provides a useful sufficient condition that leads to the satisfaction of Hörmander's condition.

Lemma 3.4. *Given a function* $u(t, x) \in C^{0,\infty}([0, T] \times \mathbb{R})$. *If for any* $(t, x) \in [0, T] \times \mathbb{R}$, *either* $u(t, x) \neq 0$ *or there exist* $n \in \mathbb{N}$ *such that* $\left|r(t)x\frac{\partial^n}{\partial x^n}(u(t, x))\right| \neq 0$, *then the SDE (3) satisfies the (one-dimensional) Hörmander's condition.*

Proof. We first rewrite SDE (3) in terms of Stratonovich differentials:

$$dX^u(t) = \left(r(t)X^u(t) + u(t, X^u(t))'\alpha(t)\right)dt - \frac{1}{2}\frac{\partial}{\partial x}u(t, X^u(t))'\sigma(t)\sigma(t)'u(t, X^u(t))dt$$
$$+u(t, X^u(t))'\sigma(t) \circ dW(t).$$

Fix $(t, x) \in [0, T] \times \mathbb{R}$. Firstly, if $|u(t, x)| \neq 0$, there exists some $i > 0$ such that $V_i \neq 0$ because of the non-degeneracy of $\sigma(t)$, and hence $\mathcal{V}_0$ contains a non-zero number.

On the other hand, if $|u(t, x)| = 0$ and $n > 0$ is the first number such that $\left|r(t)x\frac{\partial^n}{\partial x^n}(u(t, x))\right| \neq 0$ and $\left|r(t)x\frac{\partial^k}{\partial x^k}(u(t, x))\right| = 0$ for all $k < n$. Clearly, $r(t)x \neq 0$, and so $V_0 \neq 0$ and $|\frac{\partial^k}{\partial x^k}(u(t, x))| = 0$ for all $k < n$; therefore, $|\frac{\partial^k}{\partial x^k}(u(t, x)'\sigma(t))| = 0$, that is $\frac{\partial^k V_i}{\partial x^k}(t, x) = 0$ for all $k < n$ and $i = 1, \ldots, d$. As a result, for all $k < n$, $\mathcal{V}_k$ contains only 0. Since $|\frac{\partial^n}{\partial x^n}(u(t, x)'\sigma(t))| \neq 0$, there exists an i such that $\frac{\partial^n V_i}{\partial x^n}(t, x) \neq 0$. Define $U_0 := V_i \in \mathcal{V}_0$, $U_{k+1} := [U_k, V_0] \in \mathcal{V}_{k+1}$. By induction, we have $U_k = \frac{\partial^k V_i}{\partial x^k}(t, x)V_0^k + \sum_{j=0}^{k-1} F_n(t, x)\frac{\partial^j V_i}{\partial x^j}(t, x)$ for some functions F_j. Hence, $U_n = \frac{\partial^n V_i}{\partial x^n}(t, x)V_0^n + \sum_{j=0}^{n-1} F_j(t, x)\frac{\partial^j V_i}{\partial x^j}(t, x) = \frac{\partial^n V_i}{\partial x^n}(t, x)V_0^n \neq 0$. $\square$

Denote the probability density of $X^u(t)$, if exists, by $p^u(x,t)$ (i.e. $\mathbb{P}(X^u(t) < y) = \int_{-\infty}^{y} p^u(x,t)dx$ and $m^u(dx,t) = p^u(x,t)dx$). Recall a standard result:

Lemma 3.5. p^u *is the probability density of X^u if and only if p^u is the solution of the following Fokker-Planck Equation:*

$$\frac{\partial p^u(x,t)}{\partial t} + \frac{\partial}{\partial x}\left(p^u(x,t)(r(t)x + u(t,x)'\alpha(t))\right) - \frac{1}{2}\frac{\partial^2}{\partial x^2}\left(p^u(x,t)u(t,x)'\sigma(t)\sigma(t)'u(t,x)\right) = 0,$$

(19)

where $\quad p^u(x,0) = \delta_{x_0}(x).$

In light of the above lemma together with Theorem 3.1, we have the following corollary which states that if the optimal wealth process has a transition density, the necessary condition for optimality becomes the solution of a couple of HJB and Fokker-Plank equations.

Corollary 3.1. *Given an admissible control $\hat{u}$. Consider ϕ as defined by equation (6) with $\hat{m}(dx,t) = \hat{p}(x,t)dx$. If $\hat{u}$ is the optimal control and ϕ satisfies Assumption 3.1, they satisfy the following HJB equation:*

$$-\frac{\partial\phi(x,t)}{\partial t} - r(t)x\frac{\partial\phi(x,t)}{\partial x} = \max_{u(t,x)}\left[\frac{\partial\phi(x,t)}{\partial x}\alpha(t)'u(t,x) + \frac{1}{2}\frac{\partial^2\phi(x,t)}{\partial x^2}u(t,x)'\sigma(t)\sigma(t)'u(t,x)\right]$$

where $\quad \phi(x,T) = x - \gamma(x_0)x^2 + 2\gamma(x_0)x\int_{-\infty}^{\infty}\hat{p}(\xi,T)\xi d\xi,$

and the maximum in the RHS of (20) is attained at $\hat{u}$; while $\hat{p}$ also satisfies the Fokker-Planck equation:

$$\frac{\partial\hat{p}(x,t)}{\partial t} + \frac{\partial}{\partial x}\left(\hat{p}(x,t)(r(t)x + \hat{u}(t,x)'\alpha(t))\right) - \frac{1}{2}\frac{\partial^2}{\partial x^2}\left(\hat{p}(x,t)\hat{u}(t,x)'\sigma(t)\sigma(t)'\hat{u}(t,x)\right) = 0$$

where $\quad \hat{p}(x,0) = \delta_{x_0}(x).$

4. Solution to the Original Unconstrained Problem

We look for a solution in the form:

(20) $$\phi(x,t) = -\frac{1}{2}P(t)x^2 + S(t)x + s(t),$$

with $P(t) > 0$. By using (16), we can write

$$\hat{u}(x,t) = -(\sigma(t)\sigma(t)')^{-1}\frac{\frac{\partial\phi(x,t)}{\partial x}}{\frac{\partial^2\phi(x,t)}{\partial x^2}}\alpha(t)$$

(21) $$= -\left(x - \frac{S(t)}{P(t)}\right)(\sigma(t)\sigma(t)')^{-1}\alpha(t).$$

Set $q(t) = \mathbb{E}[X^{\hat{u}}(t)] = \int_{-\infty}^{\infty} \xi \hat{m}(d\xi, t)$. Substitute (20) and (21) into (17), we get

$$\frac{1}{2}\dot{P}(t)x^2 - \dot{S}(t)x - \dot{s}(t) - r(t)x(-P(t)x + S(t)) = -\frac{1}{2}\alpha(t)'(\sigma(t)\sigma(t)')^{-1}\alpha(t)\frac{(-P(t)x + S(t))^2}{-P(t)}$$

and $$-\frac{1}{2}P(T)x^2 + S(T)x + s(T) = x - \gamma(x_0)x^2 + 2\gamma(x_0)q(T)x.$$

By comparing coefficients, we further obtain the following ODE system:

(22)
$$\begin{cases} \frac{1}{2}\dot{P}(t) + r(t)P(t) = \frac{1}{2}\alpha(t)'(\sigma(t)\sigma(t)')^{-1}\alpha(t)P(t) \\ P(T) = 2\gamma(x_0). \end{cases}$$

(23)
$$\begin{cases} -\dot{S}(t) - r(t)S(t) = -\alpha(t)'(\sigma(t)\sigma(t)')^{-1}\alpha(t)S(t) \\ S(T) = 1 + 2\gamma(x_0)q(T). \end{cases}$$

(24)
$$\begin{cases} -\dot{s}(t) = \frac{1}{2}\alpha(t)'(\sigma(t)\sigma(t)')^{-1}\alpha(t)\frac{S(t)^2}{P(t)} \\ s(T) = 0. \end{cases}$$

Solving for the equations (22)-(23), we have

$$P(t) = 2\gamma(x_0)\exp\left[\int_t^T \left(2r(s) - \alpha(s)'(\sigma(s)\sigma(s)')^{-1}\alpha(s)\right)ds\right]$$

$$S(t) = (1 + 2\gamma(x_0)q(T))\exp\left[\int_t^T \left(r(s) - \alpha(s)'(\sigma(s)\sigma(s)')^{-1}\alpha(s)\right)ds\right].$$

Note that $P(t) > 0$ as required. On the other hand, since $\hat{m}$ satisfies Kolmogorov's Foward Equation in (13), we have

$$q(t + h) - q(t) = \int_{-\infty}^{\infty} x\hat{m}(dx, t + h) - \int_{-\infty}^{\infty} x\hat{m}(dx, t)$$

$$= \int_t^{t+h}\int_{-\infty}^{\infty} [r(s)x + \hat{u}(s, x)'\alpha(s)]\hat{m}(dx, s)ds.$$

Differentiating both sides, we further obtain:

$$\dot{q}(t) = \int_{-\infty}^{\infty} [r(t)x + \hat{u}(t, x)'\alpha(t)]\hat{m}(dx, t)$$

$$= \int_{-\infty}^{\infty}\left[r(t)x - \left(x - \frac{S(t)}{P(t)}\right)\alpha(t)'(\sigma(t)\sigma(t)')^{-1}\alpha(t)\right]\hat{m}(dx, t)$$

$$= \left(r(t) - \alpha(t)'(\sigma(t)\sigma(t)')^{-1}\alpha(t)\right)q(t) + \frac{S(t)}{P(t)}\alpha(t)'(\sigma(t)\sigma(t)')^{-1}\alpha(t).$$

Thus, we have

$$\frac{d}{dt}\left(q(t)\exp\left[\int_t^T \left(r(s) - \alpha(s)'(\sigma(s)\sigma(s)')^{-1}\alpha(s)ds\right)\right]\right)$$

$$= \frac{1 + 2\gamma(x_0)q(T)}{2\gamma(x_0)}\alpha(t)'(\sigma(t)\sigma(t)')^{-1}\alpha(t)\exp\left[-\int_t^T \alpha(s)'(\sigma(s)\sigma(s)')^{-1}\alpha(s)ds\right].$$

Integrating both sides on $[0, T]$ with $q(0) = x_0$, we have:

$$q(T) - x_0\exp\left[\int_0^T \left(r(s) - \alpha(s)'(\sigma(s)\sigma(s)')^{-1}\alpha(s)ds\right)\right]$$

$$= \frac{1 + 2\gamma(x_0)q(T)}{2\gamma(x_0)}\left[1 - \exp\left[-\int_0^T \alpha(s)'(\sigma(s)\sigma(s)')^{-1}\alpha(s)ds\right]\right]$$

Denote $k = e^{\int_0^T r(s)ds}x_0 + \frac{1}{2\gamma(x_0)}e^{\int_0^T \alpha(s)'(\sigma(s)\sigma(s)')^{-1}\alpha(s)ds}$. Solving for the above algebraic equation for $q(T)$, we have

$$\mathbb{E}\left[X^{\hat{u}}(T)\right] = q(T)$$

$$= x_0\exp\left[\int_0^T r(s)ds\right] + \frac{1}{2\gamma(x_0)}\left(\exp\left[\int_0^T \alpha(s)'(\sigma(s)\sigma(s)')^{-1}\alpha(s)ds\right] - 1\right)$$

$$= k - \frac{1}{2\gamma(x_0)}.$$

The optimal control is given by

$$\hat{u}(x, t) = -\left(x - e^{\int_0^t r(s)ds}\left[x_0 + \frac{1}{2\gamma(x_0)}e^{-\int_0^T (r(s) - \alpha(s)'(\sigma(s)\sigma(s)')^{-1}\alpha(s))ds}\right]\right)(\sigma(t)\sigma(t)')^{-1}\alpha(t)$$

$$(25) \quad = -\left(x - e^{-\int_t^T r(s)ds}k\right)(\sigma(t)\sigma(t)')^{-1}\alpha(t).$$

Remark 4.1. Moreover, in light of Lemma 3.4, the density of $\hat{m}$ is shown to exist due to the non-degeneracy of σ and the affine form of $\hat{u}$.

The adjoint function is given by $\phi(x, t) = -\frac{1}{2}P(t)x^2 + S(t)x + s(t)$ where

$$P(t) = 2\gamma(x_0)\exp\left[\int_t^T \left(2r(s) - \alpha(s)'(\sigma(s)\sigma(s)')^{-1}\alpha(s)\right)ds\right]$$

$$S(t) = 2\gamma(x_0)ke^{\int_t^T (r(s) - \alpha(s)'(\sigma(s)\sigma(s)')^{-1}\alpha(s))ds}$$

$$= \left(2\gamma(x_0)x_0e^{\int_0^T r(s)ds} + e^{\int_0^T (\alpha(s)'(\sigma(s)\sigma(s)')^{-1}\alpha(s))ds}\right)e^{\int_t^T (r(s) - \alpha(s)'(\sigma(s)\sigma(s)')^{-1}\alpha(s))ds}$$

$$s(t) = \gamma(x_0)k^2\left(1 - e^{-\int_t^T \alpha(s)'(\sigma(s)\sigma(s)')^{-1}\alpha(s)ds}\right)$$

$$= \frac{1}{2}\frac{\left(2\gamma(x_0)x_0e^{\int_0^T r(s)ds} + e^{\int_0^T (\alpha(s)'(\sigma(s)\sigma(s)')^{-1}\alpha(s))ds}\right)^2}{2\gamma(x_0)}\left(1 - e^{-\int_t^T \alpha(s)'(\sigma(s)\sigma(s)')^{-1}\alpha(s)ds}\right).$$

Obviously, the adjoint function satisfies Assumption 3.1, because $\dot{P}(t), \dot{S}(t)$ and $\dot{s}(t)$ are uniformly bounded in t and $X^{\hat{u}}$ satisfies Lemma 3.1.

5. Comparison with Dynamic Programming Method

In the Markowitz problem, we first add an additional constraint that $\mathbb{E}[X^u(T)] = m$, and we want to minimize

$$Var[X^u(T)] = \mathbb{E}\left[X^u(T)^2\right] - m^2.$$

Therefore the problem is equivalent to

(26) $$\min \mathbb{E}\left[X^u(T)^2\right]$$

subject to $\mathbb{E}[X^u(T)] = m$. By the method of Lagrangian multiplier, the constrained optimization problem (26) is equivalent to the following unconstrained problem:

$$\min \left\{\mathbb{E}\left[X^u(T)^2\right] - \lambda \mathbb{E}[X^u(T)]\right\},$$

for a Lagrangian multiplier $\lambda \in \mathbb{R}$, or equivalently,

(27) $$\max \left\{\lambda \mathbb{E}[X^u(T)] - \mathbb{E}\left[X^u(T)^2\right]\right\},$$

for $\lambda \in \mathbb{R}$. That means: $\hat{u}$ minimizes (26) among all admissible controls such that $\mathbb{E}[X^u(T)] = m$ if and only if there exist $\lambda \in \mathbb{R}$ such that $\hat{u}$ maximizes (27).

We have the HJB equation for value function V for the unconstrained optimization problem (27):

(28) $$-\frac{\partial V(x, t)}{\partial t} - r(t)x\frac{\partial V(x, t)}{\partial x} = \max_u \left[\frac{\partial V(x, t)}{\partial x}\alpha(t)'u + \frac{1}{2}\frac{\partial^2 V(x, t)}{\partial x^2}u'\sigma(t)\sigma(t)'u\right]$$
$$V(x, T) = \lambda x - x^2,$$

By the verification theorem for the stochastic control problem, the maximizer of the RHS in HJB equation (28) is the optimal control for the unconstrained optimization (27). If $\hat{u}$ is the maximizer, we can write

(29) $$\hat{u}(x, t) = -(\sigma(t)\sigma(t)')^{-1}\frac{\frac{\partial V(x,t)}{\partial x}}{\frac{\partial^2 V(x,t)}{\partial x^2}}\alpha(t).$$

Then the equation (28) becomes:

(30) $$-\frac{\partial V(x, t)}{\partial t} - r(t)x\frac{\partial V(x, t)}{\partial x} = -\frac{1}{2}\alpha(t)'(\sigma(t)\sigma(t)')^{-1}\alpha(t)\frac{\left(\frac{\partial V(x,t)}{\partial x}\right)^2}{\frac{\partial^2 V(x,t)}{\partial x^2}}.$$

We again set $V(x, t) := -\frac{1}{2}A(t)x^2 + B(t)x + c(t)$ to satisfy (30), we then have the following ODE system:

(31)
$$\begin{cases} \frac{1}{2}\dot{A}(t) + r(t)A(t) = \frac{1}{2}\alpha(t)'(\sigma(t)\sigma(t)')^{-1}\alpha(t)A(t) \\ A(T) = 2. \end{cases}$$

(32)
$$\begin{cases} -\dot{B}(t) - r(t)B(t) = -\alpha(t)'(\sigma(t)\sigma(t)')^{-1}\alpha(t)B(t) \\ B(T) = \lambda. \end{cases}$$

(33)
$$\begin{cases} -\dot{c}(t) = \frac{1}{2}\alpha(t)'(\sigma(t)\sigma(t)')^{-1}\alpha(t)\frac{B(t)^2}{A(t)} \\ c(T) = 0. \end{cases}$$

By solving for equations (31)-(32), we have

$$A(t) = 2\exp\left[\int_t^T \left(2r(s) - \alpha(s)'(\sigma(s)\sigma(s)')^{-1}\alpha(s)\right)ds\right]$$

$$B(t) = \lambda\exp\left[\int_t^T \left(r(s) - \alpha(s)'(\sigma(s)\sigma(s)')^{-1}\alpha(s)\right)ds\right].$$

Then (33) becomes

$$\dot{c}(t) = \frac{\lambda^2}{4}\alpha(t)'(\sigma(t)\sigma(t)')^{-1}\alpha(t)\exp\left[-\int_t^T \alpha(s)'(\sigma(s)\sigma(s)')^{-1}\alpha(s)ds\right],$$

thus, we have

$$c(t) = \frac{\lambda^2}{4}\left[1 - \exp\left[-\int_t^T \alpha(s)'(\sigma(s)\sigma(s)')^{-1}\alpha(s)ds\right]\right].$$

And the correponding optimal control is:

(34)
$$\hat{u}(x, t) = -\left(x - \frac{\lambda}{2}\exp\left[-\int_t^T r(s)ds\right]\right)(\sigma(t)\sigma(t)')^{-1}\alpha(t).$$

Hence the optimal wealth process satisfies the following dynamics:

$$dX^{\hat{u}}(t)$$
$$= r(t)X^{\hat{u}}dt - \left(X^{\hat{u}}(t) - \frac{\lambda}{2}\exp\left[-\int_t^T r(s)ds\right]\right)\alpha(t)'(\sigma(t)\sigma(t)')^{-1}(\alpha(t)dt + \sigma(t)dW(t)).$$

Hence,

$$\frac{d}{dt}\mathbb{E}\left[X^{\hat{u}}(t)\right]$$
$$= \left(r(t) - \alpha(t)'(\sigma(t)\sigma(t)')^{-1}\alpha(t)\right)\mathbb{E}\left[X^{\hat{u}}(t)\right] + \frac{\lambda}{2}\exp\left[-\int_t^T r(s)ds\right]\alpha(t)'(\sigma(t)\sigma(t)')^{-1}\alpha(t).$$

Then,

$$\frac{d}{dt}\left(\exp\left[\int_t^T \left(r(s) - \alpha(s)'(\sigma(s)\sigma(s)')^{-1}\alpha(s)\right)ds\right]\mathbb{E}\left[X^{\hat{u}}(t)\right]\right)$$

$$= \frac{\lambda}{2}\exp\left[-\int_t^T \alpha(s)'(\sigma(s)\sigma(s)')^{-1}\alpha(s)ds\right]\alpha(t)'(\sigma(t)\sigma(t)')^{-1}\alpha(t).$$

We have

$$\mathbb{E}\left[X^{\hat{u}}(T)\right] = \exp\left[\int_0^T \left(r(t) - \alpha(t)'(\sigma(t)\sigma(t)')^{-1}\alpha(t)\right)dt\right]x_0$$

$$+ \frac{\lambda}{2}\left[1 - \exp\left[-\int_0^T \alpha(t)'(\sigma(t)\sigma(t)')^{-1}\alpha(t)dt\right]\right].$$

Finally, λ can be obtained by solving the following equation:

$$m = \exp\left[\int_0^T \left(r(t) - \alpha(t)'(\sigma(t)\sigma(t)')^{-1}\alpha(t)\right)dt\right]x_0$$

(35)

$$+ \frac{\lambda}{2}\left[1 - \exp\left[-\int_0^T \alpha(t)'(\sigma(t)\sigma(t)')^{-1}\alpha(t)dt\right]\right].$$

Clearly from (35), we now know how λ depends on m, and vice versa. The corresponding variance as a function of λ is given by

$$Var[X^{\hat{u}}(T)]$$
$$= -\left(\lambda\mathbb{E}\left[X^{\hat{u}}(T)\right] - \mathbb{E}\left[X^{\hat{u}}(T)^2\right]\right) + \mathbb{E}\left[X^{\hat{u}}(T)\right]\left(\lambda - \mathbb{E}\left[X^{\hat{u}}(T)\right]\right)$$
$$= -V(0, x_0) + m(\lambda - m)$$
$$= e^{\int_0^T (2r(s)-\alpha(s)'(\sigma(s)\sigma(s)')^{-1}\alpha(s))ds}x_0^2 - \lambda e^{\int_0^T (r(s)-\alpha(s)'(\sigma(s)\sigma(s)')^{-1}\alpha(s))ds}x_0$$
$$- \frac{\lambda^2}{4}\left(1 - e^{-\int_0^T \alpha(s)'(\sigma(s)\sigma(s)')^{-1}\alpha(s)ds}\right) + \frac{\lambda^2}{4}\left(1 - e^{-\int_0^T 2\alpha(s)'(\sigma(s)\sigma(s)')^{-1}\alpha(s)ds}\right)$$
$$+ \lambda e^{\int_0^T (r(s)-2\alpha(s)'(\sigma(s)\sigma(s)')^{-1}\alpha(s))ds}x_0 - e^{\int_0^T (2r(s)-2\alpha(s)'(\sigma(s)\sigma(s)')^{-1}\alpha(s))ds}x_0^2$$
$$= \left(1 - e^{-\int_0^T \alpha(s)'(\sigma(s)\sigma(s)')^{-1}\alpha(s)ds}\right)\left(e^{-\int_0^T \alpha(s)'(\sigma(s)\sigma(s)')^{-1}\alpha(s)ds}\frac{\lambda^2}{4}\right.$$
$$\left. - \lambda e^{\int_0^T (r(s)-\alpha(s)'(\sigma(s)\sigma(s)')^{-1}\alpha(s))ds}x_0 + e^{\int_0^T (2r(s)-\alpha(s)'(\sigma(s)\sigma(s)')^{-1}\alpha(s))ds}x_0^2\right).$$

Let us reconsider the original optimization problem with mean-variance utility (4). Firstly, for each m, we can obtain the efficient portfolio by tackling the constrained problem (26); by solving for the optimal variance at each level of mean return, we can construct the efficient frontier. It is obvious that for any value of $\gamma(x_0)$, the optimal control for (4) has to be the efficient portolio for a certain value of m.

Hence, we can obtain the optimal control to mean-variance problem (4) from the set of efficient porfolio by selecting the optimal value of m so that the mean and variance of the corresponding efficient portfolio can maximize the mean-variance utility among all values of m.

Now, the mean-variance utility obtained from the efficient portfolio given m (in term of λ) becomes:

$$e^{\int_0^T \left(r(t)-\alpha(t)'(\sigma(t)\sigma(t)')^{-1}\alpha(t)\right)dt}x_0 + \left(1 - e^{-\int_0^T \alpha(s)'(\sigma(s)\sigma(s)')^{-1}\alpha(s)ds}\right)$$

$$\cdot \left[\frac{\lambda}{2} - \gamma(x_0)\left(e^{-\int_0^T \alpha(s)'(\sigma(s)\sigma(s)')^{-1}\alpha(s)ds}\frac{\lambda^2}{4} - e^{\int_0^T \left(r(s)-\alpha(s)'(\sigma(s)\sigma(s)')^{-1}\alpha(s)\right)ds}x_0\lambda\right.\right.$$

$$(36) \qquad \left.\left. + e^{\int_0^T \left(2r(s)-\alpha(s)'(\sigma(s)\sigma(s)')^{-1}\alpha(s)\right)ds}x_0^2\right)\right].$$

To maximize the mean-variance utility, we choose the best m (or equivalently, λ) so that (36) is maximized (36); indeed, the optimal λ is:

$$(37) \qquad \lambda = 2e^{\int_0^T r(s)ds}x_0 + \frac{1}{\gamma(x_0)}e^{\int_0^T \alpha(s)'(\sigma(s)\sigma(s)')^{-1}\alpha(s)ds} = 2k.$$

If we put (37) back into (34), we actually get the same optimal control as (25).

6. Conclusion

In this paper, we obtained the pre-commitment solution of continuous time mean-variance portfolio selection problem via mean-field approach, it is different from the embedding technique, which is commonly applied for solving mean-variance problem in the literature. In our work, we set the set of all feasible controls to be that of all Lipschitz continous controls. We shall explore the same problem in the future work with respect to an extended family of admissible controls so that all $\mathcal{L}^2$-integrable controls could be included in the feasible set.

Acknowledgement

The first author acknowledges the financial support from WCU (World Class University) program through the National Research Foundation of Korea funded by the Ministry of Education, Science and Technology (R31 - 20007) and The Hong Kong RGC GRF 500111. The third author-Phillip Yam acknowledges the financial support from The Hong Kong RGC GRF 404012 with the project title: Advanced Topics In Multivariate Risk Management In Finance And Insurance, The Chinese University of Hong Kong Direct Grants 2010/2011 Project ID: 2060422 and 2011/2012 Project ID: 2060444. Phillip Yam also expresses his sincere gratitude to the hospitality of both Hausdorff Center for Mathematics of the University of Bonn for his attendance at the workshop "Stochastic Dynamics in Economics and Finance".

References

1. Andersson, D. and Diehiche, B., 2010. A Maximum Principle for SDEs of Mean-Field Type. Applied Mathematics and Optimization 63(3), 341–356.
2. Bensoussan, A., Frehse, J. and Yam, S. C. P., 2013. Mean-field Games and Mean-field Type Control Theory. To be published by Springer Verlag.
3. Bensoussan, A., Sung, K. C. J., Yam, S. C. P. and Yung, S. P., 2011. Linear Quadratic Mean Field Games. Submitted.
4. Bensoussan, A., Sung, K. C. J. and Yam, S. C. P., 2012. Linear Quadratic Time-Inconsistent Mean Field Games. To appear in Dynamic Games and Applications: Special Issue on Mean Field Games.
5. Bensoussan, A., Wong, K. C., Yam, S. C. P. and Yung, S. P., 2013. Time consistent porfolio selection under short-selling prohibition: from discrete to continuous setting. To appear in SIAM J. Financial Mathematics.
6. Björk, T. and Murgoci, A., 2010. A general theory of Markovian time inconsistent stochastic control problems. Working Paper, Stockholm School of Economics.
7. Bouleau, N. and Hirsch, F., 1991. Dirichlet Forms and Analysis on Wiener Space. Walter de Gruyter, Berlin/New York.
8. Buckdahn, R., Djehiche, B. and Li, J., 2011. A general stochastic maximum principle for SDEs of mean-field type. Applied Mathematics and Optimization 64, 197–216.
9. Caines, P. E., Huang, M. and Malhame, R. P., 2007. An Invariance Principle in Large Population Stochastic Dynamic Games. Journal of Systems Science and Complexity 20(2), 162–172.
10. Caines, P. E., Huang, M., Malhame, R. P. and Nourian, M., 2012. Nash, social and centralized solutions to consensus problems via mean field control theory. IEEE Transaction on Automatic Control. To appear.
11. Chen, P., Yang, H. and Yin, G., 2008. Markowitzs mean-variance asset-liability management with regime switching: A continuous-time model. Insurance: Mathematics and Economics 43(3), 456–465.
12. Chiu, M. C. and Li, D., 2006. Asset and liability management under a continuous-time mean-variance optimization framework. Insurance: Mathematics and Economics 39(3), 330–355.
13. Denis, B. R., 2006. The Malliavin calculus. Mineola, NY: Dover Publications Inc. pp. 113.
14. Hairer, M., 2011. On Malliavin's proof of Hörmander's theorem. Bulletin des Sciences Mathématiques 135, 650–666.
15. Hörmander, L., 1967. Hypoelliptic second order differential equations. Acta Math. 119, 147171. doi:10.1007/BF02392081
16. Lasry, J. M. and Lions, P. L., 2007. Mean field games. Japanese Journal of Mathematics 2(1), 229–260.
17. Li, D. and Ng, W. L., 2000. Optimal dynamic portfolio selection: Multiperiod mean-variance formulation. Mathematical Finance 10(3), 387–406.
18. Li, D. and Zhou. X. Y., 2000. Continuous-time mean-variance portfolio selection: A stochastic LQ framework. Applied Mathematics and Optimization 42(1), 19–33.
19. A. E. B. Lim and X. Y. Zhou, 2002. Mean-variance portfolio selection with random coefficients in a complete market. Mathematics of Operations Research, 27, 101–120.
20. Markowitz, H., 1952. Portfolio selection. Journal of Finance 7(1), 77–91.

21. Merton, R. C., 1971. Optimum consumption and portfolio rules in a continuous-time model. Journal of Economic Theory 3(4), 373–413.
22. Meyer-Brandis, T., Oksendal, B. and Zhou, X. Y., 2012. A mean-field stochastic maximum principle via Malliavin calculus. Stochastics: An International Journal of Probability and Stochastic Processes, 84(5-6), 643–666.
23. Nualart, D., 2009. Malliavin calculus and its applications. CBMS Regional Conference Series in Mathematics. American Mathematical Society. 85p.
24. Pliska, S. R., 1986. A stochastic calculus model of continuous trading: optimal portfolios. Mathematics of Operations Research 11, 371–384.
25. Peleg, B. and Yaari, M. E., 1973. On the existence of a consistent course of action when tastes are changing. Review of Economic Studies 40(3), 391–401.
26. Stroock, D., 2008. An Introduction to Partial Differential Equations for Probabilists. Cambridge University Press. 215p.
27. Wei, J., Wong, K. C., Yam, S. C. P. and Yung, S. P., 2013. Markowitz's mean-variance asset-liability management with regime switching: A time-consistent approach. To appear in Insurance: Mathematics and Economics.
28. Yong, J. M., 2012. A linear-quadratic optimal control problem for mean-field stochastic differential equations. Working paper.
29. Yin, G. and Zhou, X. Y., 2003. Markowitzs mean-variance portfolio selection with regime switching: A continuous-time model. SIAM Journal on Control and Optimization 42(4), 1466–1482.